Jewish Tradition and the Challe

Jewish Tradition and the Challenge of Darwinism

EDITED BY GEOFFREY CANTOR AND MARC SWETLITZ

The University of Chicago Press
Chicago and London

Geoffrey Cantor is professor of the history of science at the University of Leeds.
Marc Swetlitz has taught history of science at the Massachusetts Institute of Technology and the University of Oklahoma.

The University of Chicago Press, Chicago 60637
The University of Chicago Press, Ltd., London

Printed in the United States of America

15 14 13 12 11 10 09 08 07 06   1 2 3 4 5

ISBN-13: 978-0-226-09276-8 (cloth)
ISBN-13: 978-0-09277-5 (paper)
ISBN-10: 0-226-09276-3 (cloth)
ISBN-10: 0-226-09277-1 (paper)

Library of Congress Cataloging-in-Publication Data

Jewish tradition and the challenge of Darwinism / edited by Geoffrey Cantor and Marc Swetlitz.
p. cm.
Includes bibliographical references and index.
ISBN 0-226-09276-3 (cloth : alk. paper) — ISBN 0-226-09277-1 (pbk. : alk. paper) 1. Evolution—Religious aspects—Judaism. 2. Evolution (Biology)—Religious aspects—Judaism. 3. Judaism and science. 4. Creation. 5. Earth—Age. I. Cantor, G. N., 1943– II. Swetlitz, Marc, 1958–
BM538.E8.J49 2006
296.3'4—dc22

2006017533

∞ The paper used in this publication meets the minimum requirements of the American National Standard for Information Sciences—Permanence of Paper for Printed Library Materials, ANSI Z39.48-1992.

Contents

Contributors

GEOFFREY CANTOR is professor of the history of science at the University of Leeds, UK. Among his publications are *Optics after Newton* (1983), *Michael Faraday: Sandemanian and Scientist* (1991), and, with John Hedley Brooke, *Reconstructing Nature: The Engagement of Science and Religion* (1998). This last work was based on the Gifford Lectures, which he and Professor Brooke delivered in Glasgow in 1995–96. He has served as president of the British Society for the History of Science and is currently codirector, with Professor Sally Shuttleworth, of the Science in the Nineteenth Century Periodical Project. His *Quakers, Jews, and Science* was published by Oxford University Press in 2005.

SHAI CHERRY received his doctorate in modern Jewish thought from Brandeis University (2001). His dissertation focused on Jewish responses to evolutionary theory. Since 2001, he has served as the Mellon Assistant Professor of Jewish Thought at Vanderbilt University. He is the author of "Three Twentieth-Century Jewish Responses to Evolutionary Theory," in *Aleph* 3 (2003): 247–90. His latest project, *No Word Unturned: Jewish Interpretations of the Hebrew Bible*, is to be published by the Jewish Publication Society.

RAPHAEL FALK is professor emeritus at the Hebrew University of Jerusalem. Beginning in 1953, he studied the effects of X-ray-induced mutations in *Drosophila melanogaster*, mutation loads in populations, chromosome organization in *Drosophila*, and mutations affecting development in *Drosophila*. Since 1983 he has been writing on concepts in genetics and evolution, history of genetic analysis, and genetics and society. He is on the editorial boards of *Biology and Philosophy* and *Theory and Biosciences* and has organized several conferences on genetics.

CARL FEIT is the Dr. Joseph and Rachel Ades Chair in Health Sciences, associate professor of biology, and chairperson of the Science Division of Yeshiva College.

A noted cancer research scientist, Dr. Feit came to the Biology Department of Yeshiva College in 1985 from the Laboratory for Immunodiagnosis at the Sloan-Kettering Institute for Cancer Research. He has conducted research on making monoclonal antibodies to help diagnose and treat cancer and has written extensively on the nature of antigens that appear on the cells of sarcomas. Dr. Feit did his undergraduate work at Yale University and Yeshiva University and holds a Ph.D. in microbiology and immunology from Rutgers University. He serves on the editorial board of *Cancer Investigation.* Dr. Feit is a founding member of the International Society for Science and Religion and serves on its Executive Committee. Dr. Feit is also an ordained rabbi and a talmudic scholar who has lectured on Talmud and taught classes for many years.

IRA ROBINSON is professor of Judaic studies in the Department of Religion of Concordia University, Montreal. He received his B.A. at Johns Hopkins University, his B.H.L. at Baltimore Hebrew College, an M.A. at Columbia University, and a Ph.D. in Near Eastern languages and civilizations at Harvard University. He has taught at Concordia since 1979 and served as the chair of the Department of Religion. Robinson edited *Cyrus Adler: Selected Letters* (2 vols., 1985), which won the Kenneth Smilen Award for Judaic nonfiction. He has also coedited *The Thought of Maimonides: Philosophical and Legal Studies* (1990), *An Everyday Miracle: Yiddish Culture in Montreal* (1990), *The Interaction of Scientific and Jewish Cultures* (1994), *Renewing Our Days: Montreal Jews in the Twentieth Century* (1995), which won a Toronto Jewish Book Award, and *Juifs et Canadiens Français dans la Société Québécoise* (2000). He has published *Moses Cordovero's Introduction to Kabbalah: An Annotated Translation of His Or Ne'erve* (1994). His most recent publication is *Not Written in Stone: Canadian Jews, Constitutions, and Constitutionalism in Canada* (2003), which he coedited with Daniel Elazar and Michael Brown.

RENA SELYA is a historian focusing on the history of the life sciences in the nineteenth and twentieth centuries, with an emphasis on evolution, genetics, and molecular biology. She is currently a visiting scholar at UCLA. Her dissertation, "Salvador Luria's Unfinished Experiment: The Public Life of a Biologist in a Cold War Democracy," is being revised for publication. She is the author of several book reviews in the *Journal of the History of Biology*, the *Journal of Interdisciplinary History*, the *Quarterly Review of Biology*, and *Nature.*

MARC SWETLITZ holds a Ph.D. in the history of science from the University of Chicago. He has taught history of science at the Massachusetts Institute of Technology and the University of Oklahoma, and was a Dibner Visiting Scholar at Brandeis University. His publications include "American Jewish Responses to Darwin and Evolutionary Theory, 1860–1890," in *Disseminating Darwinism: The Role of Place, Race, Religion, and Gender*, ed. Ronald L. Numbers and John Stenhouse (1999). He is currently a learning manager at Deloitte Services LP and continues to conduct research and write on topics related to Judaism, ecology, and evolution.

LAWRENCE TROSTER is the Jewish chaplain and associate of the Institute of Advanced Theology at Bard College in Annandale-on-Hudson, New York. He is the rabbinic fellow of the Coalition on the Environment and Jewish Life (COEJL) and the rabbinic scholar-in-residence for GreenFaith, an interfaith environmental coalition in New Jersey. Previously, he served as the rabbi of several congregations in New Jersey and Toronto. He received his B.A. from the University of Toronto and his M.A. and rabbinic ordination from the Jewish Theological Seminary of America. Rabbi Troster serves on the Interfaith Partnership for the Environment of UNEP (United Nations Environment Program) and the board of directors of the Newark School of Theology. He is also a member of the editorial board of *Conservative Judaism* and has published numerous articles on theology, environmentalism, liturgy, bioethics, and the relationship of science and religion.

RICHARD WEIKART is professor of modern European history and chair of the Department of History at California State University, Stanislaus. His dissertation, "Social Darwinism: Evolution in German Socialist Thought from Marx to Bernstein," won the biennial dissertation prize of the Forum of History of Human Science and was published in 1999. His publications include *From Darwin to Hitler: Evolutionary Ethics, Eugenics, and Racism in Germany* (2004) and articles on social Darwinism and eugenics in *German Studies Review, Journal of the History of Ideas, Isis*, and *European Legacy.*

PAUL WEINDLING is the Wellcome Trust Research Professor in the History of Medicine in the School of Humanities, Oxford Brookes University, UK. His publications include *Health, Race, and German Politics: Between National Unification and Nazism* (1989), *Epidemics and Genocide in Eastern Europe 1890–1945* (2000), and *From Medical War Crimes to Informed Consent: the Nuremberg Medical Trial and Allied War Crimes Policies 1945–50* (2004). He edited *International Health Organizations and Movements 1918–1939* (1995) and the journal *Social History of Medicine* (1992–98), and coedited *Blood and Homeland: Eugenics and Racial Nationalism in Central and Southeast Europe, 1900–1940* (2005). His research interests include international eugenics and health organizations in the twentieth century, the medical emigration to Britain in the 1930s and 1940s, and Nazi medical war crimes prosecuted at Nuremberg. He has compiled a database on medical refugees in the United Kingdom, covering over 4,800 doctors, dental surgeons, medical scientists, biologists, and nurses who came to Britain as a result of Nazism and World War II. He teaches the history of medicine, social Darwinism, and eugenics.

Acknowledgments

In the spring of 2002 we began discussing the possibility of compiling a volume of essays on Jewish responses to the Darwinian theory of evolution. Although we suspected that other scholars might be exploring various facets of the relationship among Jews, Judaism, and evolution, editing a collection of essays seemed daunting because so little had been published on this topic. This proposal gradually took shape as we corresponded with a large number of people in the fields of Jewish studies and the history of science. Gratifyingly, our correspondents encouraged us to proceed, generously offering advice and suggesting the names of potential contributors. We would like to extend our sincere thanks to all those who helped us at this early stage in the process.

We soon realized that because the topic was relatively unexplored, it would be necessary to bring the contributors together for a period of sustained discussion in order to enable them to interact constructively and exchange ideas. Norbert Samuelson and Hava Tirosh-Samuelson, both of whom have written on the relationship between Judaism and science, stepped in at this juncture and generously offered to host a conference at Arizona State University, Tempe. With impressive efficiency Hava took on the responsibilities of local organizer, and we are delighted to take this opportunity to thank her for her sterling efforts and both her and Norbert for being such gracious and enthusiastic hosts. We also thank the John Templeton Foundation for providing a grant to support the conference and the following organizations at Arizona State University for sponsoring it: the Grossman Chair of Jewish Studies, the Center for Society and Biology, the Center for the Study of Religion and Conflict, the Department of Religious Studies, and the Department of History.

The conference, entitled "Jewish Tradition and the Challenge of Evolution," was held from 28 February to 1 March 2004. In addition to the contributors, a number of experts were invited to comment on individual papers and to contribute to the discussion. We were honored by their presence and by their effort, which helped stimulate discussion and has resulted in a much-improved volume. We express our appreciation to Garland Allen, Leora Batnitzky, Simon Baumberg, Kalman Bland, John Hedley Brooke, Joel Gereboff, David Kohn, John M. Lynch, Manfred Laubicher, Jane Maienschein, Ronald L. Numbers, Michael Ruse, Norbert Samuelson, and Hava Tirosh-Samuelson for their important contributions.

This volume has also benefited considerably from the advice of two anonymous referees. We are likewise greatly indebted to Catherine Rice, Michael Koplow, and their colleagues at the University of Chicago Press for their dedication in turning the typescript into an attractive publication.

Last, but not least, we express our immense debt to the contributors, who gave their time to write papers and respond to the many questions and comments they received during the conference and afterwards. Their interest, enthusiasm, and contributions to this volume have extended and broadened our understanding of how evolutionary theory has interacted with Jewish tradition.

Geoffrey Cantor
Marc Swetlitz

Introduction

The theory of evolution has provided a profound intellectual challenge to religion over the past century and a half. Yet most studies of the impact and significance of evolution for religion have focused on just one religion, Christianity, and more specifically on Protestantism. While not underestimating the great variety of Protestant engagements with evolution, we should be careful not to base general claims about evolution and religion on this single historical example. To gain a wider and more balanced understanding of this crucially important topic in the field of science and religion, it is necessary to determine how other faiths have reacted to scientific theories of evolution. The essays in this volume seek to supplement the sparse historical literature[1] on how other faith traditions have coped with evolution by investigating the ways in which Jews—singly and communally—have engaged evolution in a variety of different historical contexts and the roles that evolutionary theory has played in modern Jewish history.

Engagement with evolutionary theory forms an important yet underexamined dimension in modern Jewish history and thought. A main theme in postmedieval Jewish historiography has been the encounter with modernity, particularly during the Enlightenment and the subsequent emancipation of the Jews in many countries. While scholars have primarily explored the role played by the scientific study of both history and the Bible in the transformation of Judaism in the nineteenth century, only recently have they begun to direct attention to the physical and biological sciences, two areas of science

1. On non-Christian religious responses to Darwinism, see Najm A. Bezirgan, "The Islamic World," in *The Comparative Reception of Darwinism*, ed. Thomas F. Glick (Austin: University of Texas Press, 1974), 375–87; Mehmet Bayrakdar, "Al-Jahiz and the Rise of Biological Evolutionism," *Islamic Quarterly* 21 (1983): 149–55; Dermot Killingley, "Hinduism, Darwinism, and Evolution in Late Nineteenth-Century India," in *Charles Darwin's "The Origin of Species": New Interdisciplinary Essays*, ed. David Amigoni and Jeff Wallace (Manchester: University of Manchester Press, 1995), 174–202. For Roman Catholic responses, see R. Scott Appleby, "Exposing Darwin's 'Hidden Agenda': Roman Catholic Responses to Evolution," in *Disseminating Darwinism: The Role of Place, Race, Religion, and Gender*, ed. Ronald L. Numbers and John Stenhouse (Cambridge: Cambridge University Press, 1999), 173–208.

that have profoundly transformed modern life and thought.[2] The essays in this volume are intended to contribute to this growing literature.

Ever since the 1860s, leading rabbis and Jewish intellectuals have addressed the challenges raised by evolutionary theory for religious belief, theological understanding, and even the very nature of Jewish identity. However, the impact of evolution on Jewish theology remains an important area for further scholarly exploration. Religious responses can also provide insights into general concerns about assimilation, Jewish-Christian relations, and the demarcation between the various movements within world Jewry. Moreover, the study of evolution can illuminate how modern science has been recruited both as an enemy and as an ally of Jews: on the one hand, anti-Semites seeking to legitimate the inferiority of the Jewish race have evoked social Darwinist notions, while on the other Jewish intellectuals have sought to construct notions of Jewish identity by appealing to the authority of the science of evolution. The essays in this volume address two important issues in the social history of science. First, in assessing the variety of Jewish understandings of evolution and reactions to it, the social history of Darwinism is extended into relatively uncharted areas. Second, in discussing how ideas drawn from evolutionary theories were deployed by writers on race, some essays contribute to our understanding of how theories of race have been framed and deployed in highly charged political contexts.

The contributors to this volume work in several academic disciplines—history, history of science, Jewish thought, and biology—as well as the rabbinates of two branches of Judaism. They therefore bring to this multifaceted subject a variety of perspectives. While the editors are aware that this collection of ten papers can address only a very few specific topics within a much broader field of research, they intend these essays to stimulate future research into other aspects of this subject. In addition, these essays provide evidence and perspectives that can be used for drawing interreligious comparisons as well as comparisons between Jewish communities in different times and places.

This volume is divided into three parts, each addressing a different set of issues. Each part has a short introduction that explains why the topic is

2. For example, David Ruderman, *Jewish Thought and Scientific Discovery in Early Modern Europe* (New Haven: Yale University Press, 1995); Yakov Rabkin and Ira Robinson, eds., *The Interaction of Scientific and Jewish Cultures in Modern Times* (Lewiston, NY: Edwin Mellen Press, 1995); David A. Hollinger, *Science, Jews, and Secular Culture: Studies in Mid-20th-Century American Intellectual History* (Princeton: Princeton University Press, 1996). See the periodical *Aleph: Historical Studies in Science and Judaism*, which spans both earlier and later periods.

important and may have been controversial and identifies important issues that deserve further research.

Darwin's Evolutionary Theory and Its Impact

In November 1859 the London publisher John Murray issued a book by Charles Darwin entitled *On the Origin of Species by Means of Natural Selection.*[3] Although it soon attracted a number of reviews—some strongly positive, some emphatically negative, and others merely insipid[4]—few contemporaries could have foreseen its impact. Over the last century and a half it has underpinned much research in the biological and human sciences and been the dominant theory for many working scientists. But its influence on areas outside science has also been immeasurable. It has deeply affected the way we think about ourselves, it has impacted profoundly on the writings of philosophers and theologians, and it has been deployed in the service of various social and political ideologies. Evolution has also been a recurrent focus for controversy.[5]

Yet this crucially important book should be understood in its historical context. Darwin was not the first to argue that species evolved. Over the half century preceding its publication there had been a number of works, such as Lamarck's *Philosophie Zoologique* (1809) and the anonymous *Vestiges of the Natural History of Creation* (1844), that had rejected the prevalent view in earlier periods that God had created each species separately and that these

3. The literature on Darwin is vast. Two recent and important biographical studies are Janet Browne, *Charles Darwin*, 2 vols. (London: Cape, 1995–2002); Adrian Desmond and James Moore, *Darwin* (London: Michael Joseph, 1991).

4. A number of early reviews are collected in David Hull, *Darwin and His Critics: The Reception of Darwin's Theory of Evolution by the Scientific Community* (Cambridge: Harvard University Press, 1973); Alvar Ellegård, *Darwin and the General Reader: The Reception of Darwin's Theory of Evolution in the British Periodical Press, 1859–1872* (Gothenburg: Götenborgs Universitet, 1958).

5. From the immense literature on the significance of the *Origin* and its historical impact here is a small selection: Jonathan Hodge and Gregory Radick, eds., *The Cambridge Companion to Darwin* (Cambridge: Cambridge University Press, 2003); David Kohn, ed., *The Darwinian Heritage* (Princeton: Princeton University Press, 1985); Peter Bowler, *Evolution: The History of an Idea* (Berkeley: University of California Press, 1983); Bowler, *Charles Darwin: The Man and His Influence* (Oxford: Blackwell, 1990); Howard E. Gruber, *Darwin on Man: A Psychological Study of Scientific Creativity*, 2nd ed. (Chicago: University of Chicago Press, 1981); Hull, *Darwin and His Critics*; Robert J. Richards, *Darwin and the Emergence of Evolutionary Theories of Mind and Behavior* (Chicago: University of Chicago Press, 1987); Michael Ruse, *The Darwinian Revolution* (Chicago: University of Chicago Press, 1981); Robert M. Young, *Darwin's Metaphor: Nature's Place in Victorian Culture* (Cambridge: Cambridge University Press, 1985).

species had remained stable up to the present day. Moreover, by 1859 most scientists accepted on the basis of geological evidence that the earth was considerably older than the six thousand years postulated by the more conservative biblical chronologies. By the time Darwin's book was published the transformation of species was a familiar theme and one that was often associated with political radicalism and atheism.

What was so novel about Darwin's book? First, Darwin possessed an impressive reputation as a naturalist and, far from offering a hasty speculation, he underpinned his case for evolution with a wealth of research and information gleaned from many naturalists all over the world. He also applied his theory to help explain observations in such diverse fields as taxonomy, morphology, embryology, animal behavior, geology, palaeontology, and biogeography. Second, he proposed an innovative and controversial mechanism for evolution—the evolution of species *by natural selection*. Put simply, Darwin postulated that individuals in any species exhibit diversity in a range of characteristics, many of which are inherited, and that these individuals compete with each other for the scarce resources needed to survive and reproduce. If certain characteristics provide individuals with an advantage in that competition, then those characteristics are more likely to be passed to the next generation, and the species will evolve. For example, if there is insufficient food for all the members of a population of sparrows, they will compete over this limited resource. If some sparrows possess an advantage over others—perhaps the faster-flying sparrows have greater access to food—then the faster are more likely to survive than the slower. And if the biological basis for the characteristic of speed is passed on to successive generations, we can envisage the speed of our hypothetical population increasing over a number of generations. Thus the species is transformed.

While Darwin's theory was not the first to postulate evolution, it needs to be distinguished from the view that evolution is teleological—that species develop linearly towards a predetermined, final form. Rather, Darwin envisaged evolution like a branching tree, with some branches splitting and others terminating. These correspond respectively to the diversification of species and to extinction. Moreover, how an actual species evolves depends on contingent local factors, such as the availability of food and the presence of predators. This dimension of Darwin's theory had important theological implications, for it was understood by Darwin and many theologians to challenge notions of a providential God whose care and benevolence extended to each individual in creation.

Darwin's theory also challenged the argument from design, especially widespread in the writings of British and American Christian clergy in the nineteenth century. If we look at any part of the natural world, impressive evidence for its design can usually be found. A flower, for example, displays design in the symmetry of its petals and in the functioning of its reproductive cycle. Such signs of design, it was argued, must be the result of a divine designer. The physical world thus justifies the notion not just of a Creator, but of a superintelligent Creator who made every species for a purpose and fitted each species into an overall plan. Then came Darwin's theory. As some commentators noted, evolution evokes an even more impressive image of God, since he did not have to create each species separately but rather possessed the intelligence to create a system in which species progressively evolved. However, others viewed evolution very differently since if species evolve unaided, then the notion of a designer God was undermined. Indeed, Darwin's theory was widely understood as being conducive to atheism, as it appears to dispense with a providential God.[6]

The centrality of competition in the process of natural selection also raised a host of problems. Nature was a battleground: nature "red in tooth and claw," to use Tennyson's famous phrase.[7] The need for many members of a species population to die so that the most fit could survive might seem to naturalize death and destruction. How could the evil at the heart of the natural world be reconciled with the traditional view of a benevolent God? The ethical implications of evolution also entered political thought, since if competition is natural, then the domination of the weak by the strong is implicitly justified. Darwin's theory could thus be seen to rationalize cruelty and—by implication—such systems as exploitative capitalism. But, argued some, it could equally legitimate the necessity for all classes to work together in order to maximize the fitness, and therefore the survival, of society as a whole. Thus social Darwinism—the application of evolution to society as theory and/or policy—has been deployed by writers from across

6. On religious responses to Darwin, see, for example, Peter Bowler, *Reconciling Science and Religion* (Chicago: University of Chicago Press, 2001); John Durant, ed., *Darwinism and Divinity: Essays on Evolution and Religious Belief* (Oxford: Blackwell, 1985); Frederick Gregory, *Nature Lost? Natural Science and the German Theological Traditions of the Nineteenth Century* (Cambridge: Harvard University Press, 1992); David N. Livingstone, *Darwin's Forgotten Defenders: The Encounter between Evangelical Theology and Evolutionary Thought* (Grand Rapids, MI: Eerdmans, 1987); James R. Moore, *The Post-Darwinian Controversies: A Study of the Protestant Struggle to Come to Terms with Darwin in Great Britain and America, 1870–1900* (Cambridge: Cambridge University Press, 1979); Jon Roberts, *Darwinism and the Divine in America* (Madison: University of Wisconsin Press, 1988).

7. Alfred Tennyson, *In Memoriam* (London: Edward Moxon, 1850), canto LVI.

the political spectrum to provide "scientific" justification for their views about human society.[8]

In addition to the host of theological, social, and political issues raised by Darwin's theory of evolution, comparative studies have shown that reactions to the theory have often been contingent on local factors.[9] For instance, the strenuous antievolutionism displayed by Christian fundamentalists throughout much of the twentieth century has been far more prominent in many parts of the United States than it has in other countries.

Yet within the wide diversity of uses of evolution two stand out as particularly important when reflecting on Jewish history from the vantage point of the early twenty-first century. The first is the use of evolution in racial theory in Germany in order to provide "scientific" credence to anti-Semitism. Darwin's theory was generally warmly welcomed in Germany, where its main popularizer in the late 1860s was Ernst Haeckel. Although socialists and communists were quick to exploit the theory in their antireligious polemics, it also attracted more conservative and nationalistic writers. Particularly after Germany's defeat in the First World War, provoking a rising tide of nationalism, Darwin's theory became increasingly aligned with reactionary groups, like the Nazis, who argued the superiority of the Aryan race—being the most highly evolved—and the inferiority of other races, including Jews, blacks, and Gypsies. This transposition of evolution to the politics of race helped to justify anti-Semitism, and by the 1920s and '30s many scientists and doctors used their authority to advocate the racial and eugenicist policies of the Nazis. After Hitler came to power in 1933 a law was passed that led to the sterilization of some 400,000 people who possessed various (allegedly) heritable diseases: that they could not reproduce would, it was thought, result in the genetic improvement of the German population. A few years later the same rationale led to the institutionally sanctioned killing of large numbers of the physically and mentally disabled. The way was paved for the mass murder of Jews, Gypsies, and other "undesirables" in the extermination camps.[10]

8. For an overview of the extensive literature on social Darwinism, see Diana B. Paul, "Darwin, Social Darwinism, and Eugenics," in Hodge and Radick, *Cambridge Companion to Darwin*, 214–39.

9. The significance of local context is discussed by David Livingstone, "Science, Region, and Religion: The Reception of Darwinism in Princeton, Belfast, and Edinburgh," in Numbers and Stenhouse, *Disseminating Darwinism*, 7–38.

10. Paul, "Darwin, Social Darwinism, and Eugenics." See also Elazar Barkan, *The Retreat of Scientific Racism: Changing Concepts of Race in Britain and the United States between the World Wars* (Cambridge: Cambridge University Press, 1992); Benno Müller-Hill, *Murderous Science: Elimination by Scientific Selection of Jews, Gypsies, and Others in Germany, 1933–1945*, trans. George

Given the ostensible link between Darwinism, genetics, and the Holocaust, many present-day evolutionists publicly distance themselves from eugenics and from any attempt to recruit evolutionary science for political purposes. Yet mention of Darwin's theory sometimes evokes its association in the public mind—especially the Jewish mind—with the Holocaust. Moreover, although it is widely known that the German racial hygienists deployed sterilization, it should be remembered that similar programs were pursued elsewhere, including most of the Eastern European countries, three provinces in Canada, and thirty states in the U.S.[11]

The other significant shadow that lies across the recent history of Darwinism is the determined opposition by fundamentalist Christians. Although many committed Christians had earlier rejected evolution on religious grounds, antievolutionism became a major religious and political movement in America only in the 1920s. The First World War raised uncomfortable moral issues about the nature of humankind and increased the sense in Southern states that they were being eclipsed socially, economically, and intellectually by the North. Antievolution became a rallying point for conservative Christians, especially those associated with the fundamentalist movement. At the same time William Jennings Bryan grasped the potential of antievolutionism as a populist political platform, becoming the central figure in the 1925 Scopes trial that focused on the freedom of a community with strong religious commitments to choose whether Darwinian evolution should be taught in its schools. Although many of the arguments went against Bryan, Scopes was found guilty of teaching evolution in violation of legislation recently passed by the state of Tennessee.[12]

As a result of the Scopes trial antievolutionism achieved its highest public visibility in the late 1920s, but, beginning in the 1960s, it gained considerable momentum by being associated with scientific creationism. Supported by a number of scientists, this research program was accompanied by the demand that creationism and evolution should be given equal time in American schools. Moreover, although the early antievolutionists subscribed to a number of incompatible theories about animal and human origins, the trend has been away from the supernatural creation of species to the view

R. Fraser (Oxford: Oxford University Press, 1988); Robert Proctor, *Racial Hygiene: Medicine under the Nazis* (Cambridge: Harvard University Press, 1988); Richard Weikart, *From Darwin to Hitler: Evolutionary Ethics, Eugenics, and Racism in Germany* (New York: Palgrave Macmillan, 2004).

11. Paul, "Darwin," 230–31.

12. Edward J. Larson, *Summer for the Gods: The Scopes Trial and America's Continuing Debate over Science and Religion* (Cambridge: Harvard University Press, 1998).

that while each kind of creature was carried on the Ark, further species differentiation subsequently occurred by a process of microevolution.

Recently the scientific credentials of creationism have been extended by a resurgence of the argument that Darwin's theory is inadequate for explaining the full diversity and complexity of life, which can be understood only as the outcome of an intelligent design. Much discussion has focused on such complex biological systems as the flagella that propel bacterial cells. In this instance the proponents of Intelligent Design, such as Michael Behe, have argued that the flagellum could not have evolved by numerous slight modifications to the system, since if any of its component parts had been completely formed the whole system would not have functioned properly. Opponents have been quick to argue that the flagellum is not an "irreducibly complex" system and that the analogies exploited by Behe are false. While this may appear to be a disagreement among biologists, it is attracting a vast amount of public attention, especially in America, since Intelligent Design is portrayed as opposed to evolutionary naturalism and is widely perceived as supporting creationism. As with creationism, increasingly vigorous voices are now calling for Intelligent Design to be taught in schools.[13]

The significance of the above developments is twofold for appreciating Jewish attitudes to evolution. First, Christian opposition to evolution, especially visible in America, has been used as a rhetorical target by both Orthodox and progressive Jews who argue for Judaism's greater openness to evolutionary ideas. At the same time, there has been a resurgence of fundamentalism not only within Christianity, but also within both Judaism and Islam.[14] This development has certainly influenced attitudes towards evolu-

13. On creationism, see Ronald L. Numbers, *The Creationists: The Evolution of Scientific Creationism* (Berkeley: University of California Press, 1993). Recent literature on Intelligent Design includes Michael Behe, *Darwin's Black Box: The Biochemical Challenge to Evolution* (New York: Free Press, 1996), and William Dembski, *Intelligent Design: The Bridge between Science and Theology* (Downers Grove, IL: Intervarsity Press, 1999). For critical assessments of Intelligent Design, see Robert Pennock, ed., *Intelligent Design Creationism and Its Critics: Philosophical, Theological, and Scientific Perspectives* (Cambridge: MIT Press, 2001), and William A. Dembski and Michael Ruse, eds., *Debating Design: From Darwin to DNA* (Cambridge: Cambridge University Press, 2004).

14. On fundamentalism, see Karen Armstrong, *The Battle for God: Fundamentalism in Judaism, Christianity, and Islam* (London: HarperCollins, 2000); Martin E. Marty and R. Scott Appleby, eds., *Fundamentalisms and Society: Reclaiming the Sciences, the Family, and Education* (Chicago: University of Chicago Press, 1993). Antievolutionary arguments appear in popular Islamic works such as Harun Yahya, *The Evolution Deceit: The Scientific Collapse of Darwinism and Its Ideological Background* (London: Ta-Ha, 2000); Yahya is also the proprietor of the Web sites www.evolutiondeceit.com and www.darwinismrefuted.com (accessed 22 August 2005).

tion among certain Jewish communities, although the extent to which these Jews share the antievolutionary attitudes of Christian fundamentalists remains to be fully explored.

While it may be informative to draw comparisons and contrasts between Jewish responses to evolution and those of the Christian fundamentalists, it should be remembered that many Christians have distanced themselves from creationism and have responded to evolution in a far more conciliatory manner. For example, influential groups of American Presbyterians, Lutherans, Episcopalians, and Unitarians have passed resolutions supporting the teaching of evolution and opposing scientific creationism.[15] Some churches have adopted a less unequivocal—if more sophisticated—position. Thus in a message to the Pontifical Academy of Sciences in 1996 Pope John Paul II stressed that evolution should be taken seriously as a physical hypothesis that is worthy of investigation. However, he also cautioned against adopting an interpretation of evolution that "considered the mind as emerging from the forces of living matter, or as a mere epiphenomenon of this matter."[16] Even in churches that have been sympathetic to evolution, there has often been a minority of dissidents. For example, over the last few years Sir Peter Vardy, a wealthy car salesman and evangelical, has been trying to introduce creationism into several schools in the north of England. This move has provoked a large number of influential Anglicans, including several bishops, to rush to defend the teaching of evolution.[17] These examples indicate that in distancing themselves from the creationists, many Christians have sought to define and defend their views on evolution. Similarly, Jews from a variety of religious positions have had to make up their minds about evolution, drawing not only on the intellectual resources of Jewish tradition but also responding to a range of creationist and anticreationist arguments circulating in society at large.

15. These are reproduced in Philip Appleman, ed., *Darwin: A Norton Critical Edition*, 3rd ed. (New York: Norton, 2001), 529–33. A similar anticreationist resolution from the Reform movement's Central Conference of American Rabbis (1984) is reproduced on p. 529.

16 John Paul II, "Message to Pontifical Academy of Sciences: Magisterium Concerned with Question of Evolution," 22 October 1996, www.cin.org/jp2evolu.html (accessed 29 August 2005). For a range of reactions to creationism by Christians from other denominations, see Roland Mushat Frye, ed., *Is God a Creationist? The Religious Case against Creation-Science* (New York: Scribners, 1983).

17. For example, "Creating in a Void," *Economist*, 24 June 2004, 38. A year earlier the matter had been raised in the House of Lords (*Hansard*, 11 June 2003, cols. 187–90), leading one noble lord to pose the following rhetorical question: "if the Government are to allow the teaching of creationism, will they also allow the teaching of pre-Copernican astronomy, that the earth is flat and that the sun goes round the earth?"

Modern Jewish History and Evolution

Throughout the early modern period, Jews lived primarily in closed, self-governing communities and their intellectual life revolved primarily around the study of halakhah (Jewish law) and kabbalah (Jewish mysticism). There were important exceptions, such as Spinoza and a small number of Jews who studied at European universities, especially at the medical school at the University of Padua. In addition, historians have demonstrated that during this period Jews were more aware of and interested in secular learning than has previously been acknowledged. While most Jews had little interaction with non-Jews and with non-Jewish cultures, several leading rabbinic authorities in halakhah and kabbalah exhibited a keen interest in the natural sciences.[18]

The impetus for Jews to engage with science is typically traced to two factors, although there are debates about their relationship and relative importance. The first is the haskalah, a movement begun by Jewish intellectuals in mid-eighteenth century Europe inspired by Enlightenment ideals. Maskilim (the followers of the haskalah, the most famous being the German philosopher Moses Mendelssohn [1729-86]) advocated secular studies both as a way to modernize Judaism and as a way to enter mainstream European society. For example, in England a few maskilim avidly accepted the Newtonian system of ideas, while Emanuel Mendes da Costa (1717-91), who served as clerk of the Royal Society of London, became a leading authority on shells and fossils.[19] The second factor was the emancipation of Jews, which started in France in 1791, when they were granted virtually the full rights of citizenship. This created new opportunities for study and work, including participation in fields of science that required university-level education. However, emancipation came with explicit, or more often implicit, conditions: in order to become full members of society Jews were expected to surrender part of their religious

18. Ruderman, *Jewish Thought*; Ira Robinson, "Kabbala and Science in *Sefer ha-Berit*: A Modernization Strategy for Orthodox Jews," *Modern Judaism* 9 (1989): 275–88; Ira Robinson, "Judaism since 1700," in *The History of Science and Religion in the Western Tradition: an Encyclopedia*, ed. Gary B. Ferngren et al. (New York: Garland, 2000), 288–90; Hava Tirosh-Samuelson, "Judaism," in *Encyclopedia of Science and Religion*, ed. J. Wentzel van Huyssteen, 2 vols. (New York: Macmillan, 2003), 1:477–83.

19. Ira Robinson, "Hayyim Selig Slonimski and the Diffusion of Science among Russian Jewry in the Nineteenth Century," in Rabkin and Robinson, *The Interaction of Scientific and Jewish Cultures in Modern Times*, 31–48; David Ruderman, *Jewish Enlightenment in an English Key* (Princeton: Princeton University Press, 2000); Geoffrey Cantor, "The Rise and Fall of Emanuel Mendes da Costa: A Severe Case of 'the Philosophical Dropsy'?," *English Historical Review* 116 (2001): 584–603.

identity, as noted in the celebrated phrase of Count Stanislas de Clermont-Tonnerre, a radical deputy in the French Assembly: "Jews should be denied everything as a nation, but granted everything as individuals."[20] In addition, emancipation was granted at different times in different places, and in many instances Jews were allowed only partial rights. Indeed, where the struggle for emancipation was most difficult and prolonged, such as the German-speaking states of Central Europe, Jews often responded more creatively.

Responses to the Enlightenment and emancipation were very diverse. In varying degrees Jews adopted, adapted, and resisted non-Jewish ways of life, modes of thought, and forms of ritual practice. In the German-speaking countries, three movements emerged by the mid-nineteenth century that had a profound impact on modern Jewish life: Reform, neo-Orthodoxy (precursor of today's Modern Orthodoxy), and Positive-Historical Judaism (precursor of today's Conservative Jewish movement in America). All three sought an accommodation with modern life, and all valued secular learning—in education, as career choices, and as resources for furthering Jewish life and thought—but in different ways. Reform was the most accommodating, neo-Orthodoxy the least. These movements took different forms and followed different chronologies and trajectories in different countries. Religious innovation among Eastern European and Russian Jews occurred primarily within the Hasidic movement, which started in eighteenth-century Poland and Lithuania. With its focus on the personality of the rabbi and on personal piety, Hasidism was initially considered a threat to traditional Judaism and the authority of halakhah. Yet, by the mid-nineteenth century the followers of Hasidism had joined with traditional Jews in order to oppose efforts by Reform and other progressive Jews to modify or abandon such ritual practices as the dietary laws and the recital of prayers only in Hebrew. As regards secular learning, a range of views existed among the more traditional Jews. While Jewish study remained the preferred choice for adult males and while some prominent rabbis did not value science, other rabbinic leaders, such as Meir Leibush Malbim (1809-79), taught that science was indispensable to a proper understanding of Torah.[21]

20. "Debate on the Eligibility of Jews for Citizenship [23 December 1789]," in Paul Mendes-Flohr and Jehuda Reinharz, eds., *The Jew in the Modern World: A Documentary History* (Oxford: Oxford University Press, 1980), 103–5.

21. See Robinson, "Judaism since 1700"; Tirosh-Samuelson, "Judaism"; Noah H. Rosenbloom, "A Post-Enlightenment Exposition of Creationism," *Judaism* 38 (1989): 460–77; Michael A. Meyer, *Response to Modernity: A History of the Reform Movement in Judaism* (Oxford: Oxford University Press, 1988), chaps. 1 and 2. For a survey of modern Jewish history with extensive bibliographies, see Hilary L. Rubinstein et al., *The Jews in the Modern World: A History since 1750* (London: Arnold, 2002).

By the time Darwin's *Origin of Species* was published in 1859, Jewish religious life had diversified greatly. Intra-Jewish rivalry, while always a factor in Jewish communal life, intensified, and new concerns about assimilation and conversion to Christianity grew rapidly as emancipation spread throughout Europe. In this context, evolution was one among many secular ideas that Jewish thinkers needed to address, but it does not appear to have been among the most pressing. Little is known about Jewish responses in the early nineteenth century to new ideas about the history of the earth, including the study of fossils, and to ideas about the transmutation of species emerging in France and Germany. Moreover, more research is needed to illuminate the period after 1859. To date, scholars have examined in some detail the responses of seven European rabbis and maskilim to scientific theories of evolution. Surprisingly, the only one to reject outright the transmutation of species was a leading Reform rabbi, Abraham Geiger (1810–74), who appealed to several arguments commonly used in the 1860s: the moral and rational gap between humans and animals, the harmony evident in organic nature, and the apparent incapacity of natural forces to produce new species. The other six, who included some traditional Jews, either accepted various forms of evolution that were consistent with their theistic beliefs, or stated that they would be receptive to such theories when the scientific community had placed evolution on a more secure foundation. These rabbis and intellectuals used various textual and theological strategies to maintain core Jewish beliefs and, it is important to note, as none of them accepted a simplistic literal interpretation of Torah, they did not oppose evolution because of any putative conflict with Genesis.[22]

22. Michael Shai Cherry, "Creation, Evolution, and Jewish Thought," Ph.D. dissertation (Brandeis University, 2001), ch. 2, provides a systematic discussion of these seven individuals: Rabbis Elijah Benamozegh, Abraham Geiger, Malbim, Samson Raphael Hirsch, Naphtali Halevy, Vittorio Ḥayim Castiglioni, and the maskil Joseph Lev Sossnitz. Edward O. Dodson, "*Toldot Adam*: A Little-Known Chapter in the History of Darwinism," *Perspectives on Science and Christian Faith* 52 (2000): 47–54, and David Kohn and Ralph Colp, " 'A Real Curiosity': Charles Darwin Reflects on a Communication from Rabbi Naphtali Levy," *European Legacy* 1 (1996): 1716–27, provide a detailed analysis of Halevy's *Toledoth Adam* (The Generations of Man), in which he argued that evolution and Torah Judaism are fully compatible. In 1876 Halevy sent a copy of his book to Darwin, together with a letter with an enthusiastic dedication: "To the Lord, the Prince, . . . the Investigator of the generation, the bright son of the morning, Charles Darwin, long may he live!" Cited in Dodson, "Toldot Adam," 48. See also Lois Dubin, "*Pe'er ha-Adam* of Vittorio Ḥayim Castiglioni: An Italian Chapter in the History of Jewish Response to Darwin," in Rabkin and Robinson, *The Interaction of Scientific and Jewish Cultures in Modern Times*, 87–102; José Faur, "The Hebrew Species Concept and the Origin of Evolution: R. Benamozegh's Response to Darwin," *Rassegna Mensile di Israel* 63 (1997): 42–66; Lawrence

The contrast between Geiger and his contemporaries belies the simple correlation between the general level of accommodation to modern life and thought and acceptance of evolution. At the same time, it remains to be determined whether these rabbis and maskilim reflected the views of the broader Jewish communities in which they lived; more attention also needs to be paid to the broader social and cultural contexts of these responses. Historians have taken this more historically situated approach in analyzing American Jewish responses to evolution between 1860 and 1890, when Reform Jews and Jews aligned with the Positive-Historical movement constituted the majority of American Jewry. Advocates of theistic evolution, as well as those who rejected evolution altogether, were present in both groups, although the proportion of Jews who accepted the evolution of species increased over time. These responses should be understood within three simultaneous contexts: the relationship between the Jewish and Christian communities, the reality of assimilation and the fear that it would be encouraged by secularized scientific ideas, and the dynamics operating between the various movements within the Jewish community. Acceptance or rejection of evolution was not solely an intellectual response, but also reflected a more extensive dialogue over Jewish identity and concern with defining the Jew's place in American society. The positions developed in the late nineteenth century—especially by leading Reform rabbis—continued to shape discussions of evolution well into the twentieth.[23]

As the nineteenth century drew to a close, the place of Jews in the modern world and the Jewish encounter with evolution were irrevocably shaped by the emergence of anti-Semitism. While hatred of Jews has a long history, the rise of nationalism and of theories of race in the second half of the nineteenth century combined with existing antipathies to produce "anti-Semitism," a term coined by the German Wilhelm Marr in 1879. The pogroms in Russia, the exclusion of Jews from political office in Vienna, and the Dreyfus affair in 1890s France were among the more prominent episodes in the rise of anti-Semitic activity across Europe. One response to this rapidly deteriorating situation was the growth of Zionism. Anti-Semitism challenged the promise of

Kaplan, "*Torah u-Madda* in the Thought of Rabbi Samson Raphael Hirsch," *Bekhol Derakhekha Daehu* 5 (1997): 5–31; Marc Swetlitz, "Responses of American Reform Rabbis to Evolutionary Theory, 1864–1888," in Rabkin and Robinson, *The Interaction of Scientific and Jewish Cultures in Modern Times*, 103–25.

23. See Naomi W. Cohen, "The Challenges of Darwinism and Biblical Criticism to American Judaism," *Modern Judaism* 4 (1984):121–57; Marc Swetlitz, "American Jewish Responses to Darwin and Evolutionary Theory, 1860–1890," in Numbers and Stenhouse, *Disseminating Darwinism*, 209–46.

the Enlightenment and the emancipationists' hope that Jews would be welcomed as equal citizens in the countries in which they lived. Zionist leaders therefore argued that the creation of a nation (or cultural center) in the ancient homeland of Palestine would be the only way Jews could gain acceptance in the modern world. Not all Jews agreed, and many prominent rabbis, lay leaders, and intellectuals opposed Zionism and argued that Jews could, and should, strengthen Jewish life and religion wherever they lived.

Science and scientists played important roles in many aspects of this struggle. As mentioned above, theories of evolution and inheritance were used to support racial anti-Semitism and various forms of European social and political ideologies, including Nazism. However, historians continue to offer conflicting interpretations. While some view evolutionary discourse as mere rhetoric, others argue that evolutionary theory was a central component of Nazi ideology. Again, while some point out that racist social Darwinism was a mere vulgarization of science, others see racist social Darwinism intimately connected to evolutionary theory and to the social interests of those scientists who supported Nazism.[24] At the same time, evolution and science more generally were not exclusively the prerogative of the anti-Semite. Zionism itself was in part shaped by trends in secular European thought, including nationalism and racial ideas. Some Zionists viewed evolutionary theory as a conceptual framework for understanding the detrimental effects of Diaspora life and argued for the positive biological benefits that would accrue to Jews in Palestine. From the end of the nineteenth century, a number of Jewish anthropologists, statisticians, and physicians conducted research and published scientific papers on race and the "Jewish question" in order to bolster their Zionist—or anti-Zionist—views about the place of Jews in modern society. For some scientists, evolutionary theory played a key role in understanding the Jewish condition: these scientists drew from the range of available ideas, such as Darwinian natural selection, the "struggle for existence" among individuals and social groups, Mendelian genetics, and the Lamarckian inheritance of acquired characteristics. Discussions about evolutionary ideas, therefore, formed part of a larger project of adopting and adapting science to both define and defend Jewish secular identity and the place of Jews in European—and later Israeli—society.[25]

24. Paul Weindling, "Dissecting German Social Darwinism: Historicizing the Biology of the Organic State," *Science in Context* 11 (1998): 619–37; Weikart, *From Darwin to Hitler.*
25. John M. Efron, *Defenders of the Race: Jewish Doctors and Race Science in Fin-de-Siècle Europe* (New Haven: Yale University Press, 1994); Raphael Falk, "Zionism and the Biology of the Jews,"

The rise of the Nazis in Germany and the ensuing destruction of European Jewry had a profound impact both directly and indirectly on Jewish life and thought and on discussions about evolution. The concept of "race" became discredited among biologists and social scientists—Jews playing leading roles in that effort—and social and cultural explanations became prominent in the social sciences, where Jews continued to work in large numbers. Interest in human heredity continued, using new tools in human and population genetics. In Israel this research was part of an ongoing public debate, stimulated by the influx of immigrants from many countries, about the origin of the Jewish people.[26] The Holocaust also transformed the demography of world Jewry. The United States and Israel became leading centers for Jewish life and thought. Immigrant rabbis, philosophers, and theologians brought with them new ideas about religion, shaped by existentialism and phenomenology, which challenged the Enlightenment legacy of the primacy of reason and science. The increasing awareness of the enormous challenge that the Holocaust posed to Enlightenment ideas, and the increasing role that the State of Israel played in Jewish thought, provided fertile ground for the growth of these new intellectual currents. With these predominant foci, most rabbis and theologians paid little attention to science, including evolution, except for the issues raised by biomedical ethics. Only in the last decade of the twentieth century did some rabbis and Jewish intellectuals rekindle an interest in Judaism's relationship to science, including evolution, fueled in large part by a renewed interest in kabbalah throughout the Jewish world.[27]

Alongside these new developments, the opportunities and challenges of the emancipation of Jews and the Enlightenment continued to shape Jewish life and thought in the second half of the twentieth century. While Jews have been entering science in large numbers since the mid-nineteenth century, and have achieved prominence in many fields, most Jewish scientists were fairly assimilated, even secular. Rarely did they speak about their faith in

---

Science in Context 11 (1998): 587–607; Mitchell Hart, "Racial Science, Social Science, and the Politics of Jewish Assimilation," *Isis* 90 (1999): 268–97; Mitchell Hart, *Social Science and the Politics of Modern Jewish Identity* (Stanford: Stanford University Press, 2000).

26. Daniel J. Kevles, *In the Name of Eugenics: Genetics and the Uses of Human Heredity* (Berkeley: University of California Press, 1985); Barkan, *The Retreat of Scientific Racism*; Nurit Kirsh, "Population Genetics in Israel in the 1950s: The Unconscious Internalization of Ideology," *Isis* 94 (2003), 631–55.

27. Robert G. Goldy, *The Emergence of Jewish Theology in America* (Bloomington: Indiana University Press, 1990). See also Swetlitz (chapter 2) in this volume.

relation to their scientific work. At the same time, most rabbis and Jewish theologians were not well versed in science.[28] This changed after the Second World War, when Orthodox Jews began to enter scientific professions in significant numbers while maintaining their Jewish identities and practices. This trend is manifested in the formation of the Association of Orthodox Jewish Scientists in 1948. In this new social reality, questions of accommodation and resistance to secular culture achieved prominence on the agenda of the Orthodox community. These questions gained additional urgency in the last three decades of the twentieth century as a result of the "move to the right" by Modern Orthodoxy and the growth of ultra-Orthodox groups that strongly resist secular culture, the best known being the Lubavitch Hasidim. Evolution became one topic among many within a broad-ranging debate and the various strategies of accommodation or resistance to evolution formed part of the broader response by Orthodox groups, and individuals, to the non-Jewish world.[29]

However, prominent events in Israel and America have recently attracted attention to evolution and forced sections of the Jewish community to take a stand on whether evolution is compatible with Judaism. In the late 1990s a young ultra-Orthodox rabbi, Nosson Slifkin, began publishing a series of popular books that sought to integrate zoology—portrayed as the study of God's creatures—into a traditional Jewish framework. Slifkin's books and the educational programs he initiated were welcomed at first by all segments of Orthodoxy, including the ultra-Orthodox. Then, in September 2004, a number of leading ultra-Orthodox rabbis condemned his writings as heretical, including among their criticisms his support for evolution and his claim that the earth is millions of years old. While Slifkin maintains that in encompassing evolution he has been faithful to traditional sources and is merely reiterating the position of a number of traditional rabbinic authorities, such as Samson Raphael Hirsch (1808–88) and Abraham Isaac ha-Kohen Kook (1865–1935), the ultra-Orthodox community has been split over whether Slifkin's views undermine Torah and the authority of the rabbinic sages.

28. Yakov Rabkin, "The Interaction of Scientific and Judaic Cultures: An Historical Overview," in *The Interaction of Scientific and Jewish Cultures in Modern Times*, 3–30; Norbert Samuelson, "Judaism, History of Science and Religion, Modern Period," in van Huyssteen, *Encyclopedia of Science and Religion*, 1:491–96; David A. Hollinger, "Why Are Jews Preeminent in Science and Scholarship? The Veblen Thesis Reconsidered," *Aleph* 2 (2002), 145–63.

29. Samuel C. Heilman and Steven M. Cohen, *Cosmopolitans and Parochials: Modern Orthodox Jews in America* (Chicago: University of Chicago Press, 1989). See also Robinson (chapter 3), Cherry (chapter 7), and Selya (chapter 8) in this volume.

These reactions to Slifkin's writings also reflect internal divisions within ultra-Orthodoxy over how to respond to the challenge of modernity.[30]

At the same time, recent attempts by Christian fundamentalists to introduce Intelligent Design into public school science curricula have focused the wider American Jewish community's attention on evolution. Particularly significant was the high-profile case in Dover, Pennsylvania, held in the autumn of 2005, at which parents successfully challenged the requirement that a statement be read in biology classes informing students to avoid committing themselves to Darwin's theory and advising them to read up on Intelligent Design. This case attracted much publicity in the American Jewish press and motivated dozens of rabbis to deliver sermons, many given on Rosh Hashanah, on the issues it raised. Most rabbis and Jewish commentators applauded the judge's subsequent decision that the school district's actions violated the separation of church and state, and they reaffirmed the view that a naturalistic scientific explanation for the evolution of life is compatible with Jewish faith. Yet these views have not gone unchallenged. Because of growing diversity of opinion within American Jewry over government support for religion, an increasing number of rabbis and community leaders have expressed support for the teaching of Intelligent Design in public schools. Moreover, rabbis ranging from ultra-Orthodox through to Renewal are becoming progressively more concerned that the naturalistic assumptions underpinning evolutionary theory, and modern science more broadly, are conducive to secularism and so threaten Jewish faith. Thus the current controversies over evolution are exposing the broader social, political, and theological divisions within the Jewish community.[31]

30. Nosson Slifkin, *The Science of Torah: The Reflection of Torah in the Laws of Science, the Creation of the Universe, and the Development of Life* (Southfield, MI: Targum; Nanuet, NY: Feldheim, 2001); Jennie Rothenberg, "The Heresy of Nosson Slifkin: A Young Orthodox Rabbi Is Banned for His Views on Evolution," *Moment*, October 2005, 37–43, 45, 58, 70, 72. Slifkin posted a large number of documents and information related to the controversy at http://zootorah.com/controversy/controversy.html.

31. Jonathan Mark, "No Debate Over 'Intelligent Design,'" *Jewish Week* [New York] (25 March 2005), available at http://www.thejewishweek.com/news/newscontent.php3?artid=106749; E. B. Solomont, " 'Intelligent Design' Battles Rage On: Jewish Groups Get Involved as Legal Battles Spread from State to State," *Forward* (20 January 2006), http://www.forward.com/articles/7212. Entering "rabbi sermon 'intelligent design' " into Google search pulls up several pages of links to sermons posted on synagogue web sites (20 January 2006).

PART ONE

Historical Perspectives on Jewish Responses to Evolution

How have Jews responded to the challenge of evolution? Have their responses been different from those of contemporary Christians and members of other faiths? The essays in this part bring us closer to answering these basic questions. However, in line with recent historiography, these questions can be addressed only if we focus on local contexts in the following two respects. First, many Jewish communities display diversity in their social, political, and religious composition, and the relationship between such diversity and views about evolution needs to be explored. Second, the historically specific relationship between a Jewish community and the wider (often Christian) society in which it is located will need to be addressed, drawing on the writings of social historians on the place of Jews in different historical and cultural settings. Israel presents a unique context, since Jews form a majority and the Jewish population is incredibly diverse. The conception of context also requires further broadening to encompass those strongly bonded religious communities that are concentrated in a number of different geographical locations. For example, whether they are in New York, Jerusalem, or Paris, present-day Modern Orthodox Jews share similar beliefs and practices.

Another major historiographical theme that intersects our topic is the Jewish reaction to modernity. Innovative science, of which evolution is a paradigm example, usually forms a central element of the modernist worldview, and reactions to evolution are often symptomatic of a wider response to modernism, progress, and social change. Thus, those who proclaim themselves as modernists generally embrace evolution, while fundamentalists often reject evolution as antithetic to their religious traditions.

The above discussion provides a framework for investigating how Jewish communities have responded to evolution and for understanding the arguments they have proffered for or against evolution. Three such communities are examined in the following chapters. In the first, Geoffrey Cantor examines Anglo-Jewry in the second half of the nineteenth century, including the activities of Raphael Meldola, an avid Darwinian, and he analyzes its responses in the light of the social and cultural location of Jews in English society. While the Jewish community in late nineteenth-century England was relatively homogeneous, Judaism in twentieth-century America was very diverse. Marc Swetlitz and Ira Robinson therefore examine two parts of the religious spectrum. Swetlitz focuses on Reform, Conservative, and Reconstructionist rabbis, examining the ebb and flow of their interest in evolution, as well as the challenges that these Jews faced in confronting natural selection. Robinson shows that the dominant American and Israeli Orthodox responses to evolution should be understood as extensions of traditional strategies deployed by Jews in addressing science.

These three papers suggest how future research can explore a range of illuminating comparisons. For example, building on the studies of nineteenth-century responses to evolution by British Jews (Cantor, chapter 1) and by Reform rabbis in America (Swetlitz, chapter 2),[1] there is considerable scope for extending analysis to other nineteenth-century Jewish communities, especially those undergoing rapid acculturation and modernization, such as the Jewish communities of France and Germany. Likewise Robinson's study of Orthodoxy—with its principal focus on North America and Israel during the postwar period—could usefully be complemented by research on the several and diverse European Orthodox communities in the first half of the twentieth century and also by more detailed study of ultra-Orthodox communities around the world. Research could also look at national groups, for example, Russian Jewry during the decades of Communist repression or German Jews of the interwar period, when evolution was strongly ideological and overlaid with racist connotations. More generally, studies of this kind hold out the exciting prospect of drawing sophisticated comparisons between reactions to evolution in contemporary Jewish communities in different locations. Such studies will also necessarily address the relation between Jews and the majority culture and open for further investigation comparisons between the ways in which, say, contemporary Jewish, Muslim,

1. See also Marc Swetlitz, "Responses of American Reform Rabbis to Evolutionary Theory, 1864–1888," in *The Interaction of Scientific and Jewish Cultures in Modern Times*, ed. Yakov Rabkin and Ira Robinson (Lewiston, NY: Mellen, 1995), 103–25; Swetlitz, "American Jewish Responses to Darwin and Evolutionary Theory, 1860–1890," in *Disseminating Darwinism: The Role of Place, Race, Religion, and Gender*, ed. Ronald L. Numbers and John Stenhouse (Cambridge: Cambridge University Press, 1999), 209–46.

and Christian communities in a specific location responded to the challenges of evolution.

Of the papers in this section, Swetlitz's, with its focus on the changing responses to evolution by progressive Jewish thinkers in America over several decades, is the most strongly concerned with chronology. The chronological approach is particularly useful for seeing how communities responded to both internal and external pressures. For example, what impact did the Scopes trial have on the various sections of American Jewry? Again, how are Jewish Americans responding to the promotion of faith-based education by Christian evangelicals? Answering such questions would contribute significantly to American Jewish history.

A related area that deserves further exploration and highlights historical context is the role that evolution theories have played in shaping the biography and intellectual life of individual Jews. Darwin's ideas have led many individuals to think deeply about themselves and their Jewish identity, and about how to combine an evolutionist perspective with traditional religious commitments. Exposure to evolution may also have impelled some to adopt a secular worldview. Of particular historical interest are the biographies of those rabbis, philosophers, doctors, and scientists who have a deep—often professional—interest in addressing the topic of science and religion. Why, for example, have some of the most vociferous Modern Orthodox commentators on evolution been professional physicists or engineers, rather than biologists?[2]

Moreover, as Simon Baumberg reminded participants at the Phoenix conference, historians should look closely at the biographies of Jewish biologists, such as Paul Ehrlich, Vladimir (Waldemar) Haffkine, Ilya (Elie) Metchnikoff, Jacques Loeb, Stephen Jay Gould, Lynn Margulis, and Robert Pollack, in order to appreciate the roles of religious, secular, and evolutionary themes in shaping their lives. While more attention has been given to Jews who became mathematicians, physicists, or chemists, we should not neglect Jewish biologists and naturalists, who may be more numerous than has been assumed. For example, some leading critics of the modern synthesis in evolutionary biology and of sociobiology, including Stephen Jay Gould and Richard Lewontin, were Jewish, and it has been claimed that their opposition stemmed in part from concern that these fields are likely to encourage anti-Semitism because they emphasize genetic determinism and evolutionary progress, which often embed notions of racial hierarchy.[3] More research is

2. Robinson (chapter 3) and Cherry (chapter 7) have begun to explore the motivation of these writers.

3. Michael Ruse, *Mystery of Mysteries: Is Evolution a Social Construction?* (Cambridge: Harvard University Press, 1999), 144–45, 165–66, 168.

needed to understand which social factors brought Jews to biology and to other branches of science (not forgetting psychology, sociology, and anthropology) and whether their attitudes to science in general and to the application of evolutionary theory in particular were colored by their Jewish backgrounds and commitments.

1

Anglo-Jewish Responses to Evolution

Geoffrey Cantor

> It is astonishing to see the great hold upon the Cambridge school obtained by de Vries & Bateson. Even Francis Darwin is more or less under their dominion! I dare say you noticed that his Presidential Address to the Brit. Assoc. in Dublin last year was distinctly Lamarckian. But I for one feel sure that in a few years we shall see "Mutationism" defunct & "Mendelism" assigned to its proper place. Among the speeches delivered in the Senate House [E.] Ray Lankester's was the only one which really reaffirmed the true Darwin-Wallace position. . . . I cannot make out why I was not asked to contribute to the Memorial volume. I suppose I am too "Darwinian" for that school![1]

Thus Raphael Meldola wrote to the aging Alfred Russel Wallace, reporting the Darwin centenary celebrations of June 1909. Although more than thirty years younger than Wallace, and Lankester's junior by seven years, Meldola viewed himself as a surviving member of an earlier generation still holding firm to "the true Darwin-Wallace position" that adopted natural selection as the principal mechanism by which evolution operated. Yet this robust naturalistic position had generally been associated with materialism and thus irreligion. The quick-tempered Lankester provides a paradigm example, since he championed the theory of evolution by natural selection, preached materialism, and took every opportunity to set science against religion in order to denigrate both religion and religious institutions.[2] Unwilling to

For helpful comments on an earlier version of this paper I am most grateful to the participants at the Jewish Tradition and the Challenge of Evolution conference, especially John Lynch and Marc Swetlitz. For permission to quote from manuscript material I thank the British Library and Stephen Simpson, Professor of the Hope Entomological Collections at the Oxford University Museum of Natural History.

1. Raphael Meldola to Alfred Russel Wallace, 28 June 1909, British Library, Addit. MSS 46437: 146. The memorial volume mentioned is A. C. Seward, ed., *Darwin and Modern Science: Essays in Commemoration of the Centenary of Charles Darwin* (Cambridge: Cambridge University Press, 1909).
2. Peter Bowler, *Reconciling Science and Religion: The Debate in Early Twentieth-Century Britain* (Chicago: University of Chicago Press, 2001), 63–65.

adopt the guise of agnosticism—a term coined by Huxley—Lankester was one of the most outspoken atheistic scientists of the day.

Meldola's advocacy of a materialist version of evolution and his affinity with such avowedly antireligious naturalists as Lankester would seem to place him firmly in the antireligious camp. Yet, far from being opposed to religion, he was a committed, although not deeply observant, Jew. The example of Meldola brings into sharp focus the question of whether the response of the Anglo-Jewish community towards the theory of evolution differed from that of contemporary Christians and whether Jews were just as oppressed by the evident sense of dissonance and even conflict that marked much of the public discourse concerning evolution in the dominant Anglo-Christian world.[3] This paper therefore sets out to investigate how the Anglo-Jewish community responded to Darwinism during the third of a century following the publication of the *Origin of Species* in 1859. With the above questions in mind I shall examine a range of Anglo-Jewish responses to science in general and to evolution in particular. My main source will be the Jewish periodical press,[4] especially the *Jewish Chronicle* (the foremost Jewish weekly, which targeted the fairly assimilated upper and middle-class Jews). Since most of the writers whom I cite were editors, journalists, and rabbis—rather than scientists—I shall be focusing principally on an audience that possessed little expertise in science. At this time very few English Jews entered science, and of those who did Meldola, who is better known as a chemist and professor of chemistry at Finsbury Technical College, stands out as the most eminent Jewish naturalist of the period.[5] We shall return to his views on evolution in the final section. This paper therefore possesses two distinct but overlapping foci that provide two rather different narratives for assessing Anglo-Jewish responses to Darwin.

3. Alvar Ellegård, *Darwin and the General Reader: The Reception of Darwin's Theory of Evolution in the British Periodical Press, 1859–1872* (Gothenburg: Götenborgs Universitet, 1958). Although Ellegård's survey stopped in 1872, religious controversies over science in general and evolution in particular appear to have peaked following Tyndall's 1874 Belfast address, discussed below in this chapter.

4. In addition to the *Jewish Chronicle* (*JC*), my principal sources are the *Hebrew Observer* (*HO*), the *Jewish Record* (*JR*), and the *Jewish World* (*JW*).

5. The eccentric zoologist Walter Rothschild, who amassed private collections at Tring, published extensively and was principally concerned with describing and classifying specimens, but he did not address evolutionary problems. Also, he was somewhat peripheral to the scientific establishment. Miriam Rothschild, *Dear Lord Rothschild: Birds, Butterflies, and History* (London: Hutchinson, 1983); Edward J. Larson, *Evolution's Workshop: God and Science on the Galápagos Islands* (New York: Basic Books, 2001), 145.

The Anglo-Jewish Context

As Todd Endelman and David Englander have emphasized, a relative absence of social barriers enabled eighteenth- and nineteenth-century English Jews to achieve a high degree of assimilation, compared with their coreligionists in most European countries.[6] While a small and prosperous Jewish social elite existed, other sections of the mid-Victorian Anglo-Jewish community were becoming increasingly prosperous, and many upwardly mobile Jews were entering the middle classes.[7] They manifested a growing sense of social integration, often sending their sons to such traditional public schools as St. Paul's or to academically innovative establishments like University College School. An increasing number of Jewish students attended universities, especially University College London and the University of Cambridge. More Jews moved into the professions, resulting in 1891 in the formation of a society for Jewish professionals, the Maccabæans, which soon attracted over two hundred members. English Jews manifested a profound sense of Englishness and participated increasingly in the social and political life of the nation, sharing in its prosperity and sense of progress as technology and the British empire flourished. Modeling themselves on their Christian neighbors, they were—to use Englander's phrase—Anglicized, but not Anglican.[8]

Yet despite its alignment with English mores, Anglo-Jewry experienced many social tensions. Although vituperation in England was far less intense than in most other European countries, Anglo-Jewry was nevertheless subjected to a stream of anti-Semitic propaganda and to the blandishments of the conversion societies.[9] In popular literature Jews were often portrayed as morally and intellectually inferior to Christians. The Jew was caricatured as greedy and obsessed with making money; as one visitor to Houndsditch (the main Jewish area) in 1867 noted: the "organs of vision" of the Jew were

6. Todd M. Endelman, *The Jews of Georgian England 1714–1830: Tradition and Change in a Liberal Society*, 2nd ed. (Ann Arbor: University of Michigan Press, 1999); David Englander, "Anglicized Not Anglican: Jews and Judaism in Victorian Britain," in *Religion in Victorian Britain*, vol. 1, *Traditions*, ed. Gerald Parsons (Manchester: Manchester University Press, 1988), 235–73.

7. We will not be considering the new immigrants who settled in Britain in large numbers after 1881.

8. Englander, "Anglicized Not Anglican."

9. These issues are discussed in general histories of Anglo-Jewry including Geoffrey Alderman, *Modern British Jewry*, 2nd ed. (Oxford: Oxford University Press, 1992); Todd M. Endelman, *The Jews of Britain 1656–2000* (Berkeley: University of California Press, 2002); David S. Katz, *The Jews in the History of England 1485–1850* (Oxford: Oxford University Press, 1996).

thought to be "directed mammonward." The Jew, he continued, was "but a low-flying and lumbering, albeit an industrious and copiously perspiring, bird, . . . satisfied to burrow in muck and grow smugly sleek on such scraps and offal as the world and his wife overlooked, or, knowing the existence of, despised."[10] Jews were frequently portrayed as uneducated and therefore incapable of pursuing demanding subjects such as science and medicine. Indeed, as one earlier writer noted, the Jew was the antithesis of the cultured Englishman who, among other attributes, appreciated the natural world.[11] This anti-Semite thus portrayed the pursuit of science as alien to Jews.

Until the mid-1860s or possibly later, the community's contribution to science and literature was indeed minimal. Moreover, although a handful of Jewish scholars, mostly trained in Germany, had contributed to Hebrew and Jewish studies, the general level of religious learning was shamefully low. Jews' College (founded 1853), which trained students for the rabbinate, attracted little support from the community, and the number of students was small. A Jewish day school associated with the college was opened in 1855 but was forced to close in 1879 owing to low enrollment. Faced with the community's poor showing in both secular and religious studies its leaders strove to challenge anti-Semitic claims that Jews were intellectually deficient by amply demonstrating any success achieved by Jews in literature and science. Thus while chastising the community for its general failure to excel in secular subjects,[12] the Jewish press publicized the achievements of those who had gained distinction in science, medicine, and other areas. For example, in reporting that a Jewish student had obtained first-class honors in medicine at University College London, the *Jewish Chronicle* chastised Charles Dickens for his portrayal of Jews and pointed out that Jews could indeed make intellectual contributions.[13] Even stronger evidence was obtained from other countries, especially Germany, and from periods when the Jewish contribution to science was dominant. Thus, in reviewing Matthias Schleiden's book on the historical importance of medieval Jewish learning, a reviewer in the *Jewish Chronicle* noted the great successes of Jews "in the domain of science at a period when the profoundest mental darkness covered all Christendom;

10. James Greenwood, *Unsentimental Journeys; or Byways of the Modern Babylon* (London: Lock and Tyler, 1867), 163–64. Greenwood subsequently changed his view of Jews.

11. "The Jewish Slopseller," *Mirror of Literature* 9 (1827): 257–60.

12. For example, editorial in *JW*, 14 November 1873, 4. See also Geoffrey Cantor, "Sussex Hall (1845–1859) and the Revival of Learning among London Jewry," *Jewish Historical Studies* 38 (2003): 105–23; Geoffrey Cantor, *Quakers, Jews, and Science: Religious Responses to Modernity and the Sciences in Britain, 1650–1900* (Oxford: Oxford University Press, 2005), 142–47.

13. *JC*, 27 August 1852, 371.

when the fiercest fanaticism and the most woeful superstition had brutalised the masses; and when men of science were burnt at the stake as magicians."[14] These recurrent references to the intellectual successes of Jews illustrate Anglo-Jewry's sensitivity to the charge that Jews are inferior to Christians, a charge that threatened to undermine the attempts made by many Jews to align themselves with respectable English society.

To appreciate Jewish responses to science we too must adopt a comparative perspective. Prior to midcentury, the majority of Christian writers had unequivocally proclaimed that science was not only compatible with Christianity but also strengthened religious belief. The precise arguments depended on the writer's theological stance. For example, John Hedley Brooke has argued that the Cambridge polymath William Whewell considered that the study of the Creation should nudge men towards richer perceptions of their religious lives. Likewise, William Kirby, a High Anglican and the author of a Bridgewater Treatise, considered that both the Bible and the Book of Nature manifested the same divine plan. He therefore asserted that "in order rightly to understand the voice of God in nature, we ought to enter her temple with the Bible in our hands."[15] Yet by midcentury many Christians were increasingly troubled by the repeated salvos of materialists, evolutionists, and others who posed a severe threat to religious faith. By the 1860s, and especially the 1870s, an increasing number of Christians viewed religion and science as uncomfortable (even if not yet estranged) bedfellows, while others agreed with those atheists who asserted that recent developments in science contradicted religion.

We should also consider two topics that were prominent in Christian responses to evolution but rarely entered into Anglo-Jewish reflections. The first is materialism, which was roundly condemned by the vast majority of Christians.[16] In discussing this perennial topic I shall focus on John Tyndall's

14. "Judaism and Science," *JC*, 28 September 1877, 9–10. The reviewer was probably the editor, Abraham Benisch. On Schleiden see Ulrich Charpa, "Matthias Jakob Schleiden (1804–1881): The History of Jewish Interest in Science and the Methodology of Microscopic Botany," *Aleph: Historical Studies in Science and Judaism* 3 (2003): 213–46.

15. John Hedley Brooke, "Indications of a Creator: Whewell as Apologist and Priest," in *William Whewell: A Composite Portrait*, ed. Menachem Fisch and Simon Schaffer (Oxford: Clarendon Press, 1991), 149–73, on 167; William Kirby, *On the Power Wisdom and Goodness of God as Manifested in the Creation of Animals and Their History, Habits, and Instincts* (London: Pickering, 1835), 45. The Bridgewater Treatises were a series of eight books published in the 1830s that, according to the Earl of Bridgewater's bequest, were intended to demonstrate the "power, wisdom, and goodness of God, as manifested in the Creation."

16. Probably the best-known Christian materialist was the late eighteenth-century Socinian (and developer of the phlogiston theory) Joseph Priestley, who drew the fire of his Anglican opponents.

presidential address before the British Association's annual meeting held in Belfast in 1874, which, as Frank Turner has argued, was the preeminent manifestation of the conflict between religion and science in the Victorian period.[17] Utilizing this prominent platform to spread his ideas, Tyndall surveyed the history of materialism, from the ancients to the contemporary application of materialist doctrines in the fields of evolution, energy conservation, and physiological psychology. As Bernard Lightman's close study of reactions to Tyndall's address in the periodical press shows, the primary concern of Christian writers was Tyndall's advocacy of materialism. As one typical reviewer noted, Tyndall had "adopted, and would now wish to promulgate, the Religion of Materialism."[18] Tyndall's materialism eliminated the need for an immaterial soul and threatened the foundations of Christianity.

Nine days after Tyndall's address the *Jewish Chronicle* devoted an editorial to the subject. Remarkably the editor, Michael Henry, made no mention of materialism but instead dwelled on several other issues (which will be discussed below).[19] With one possible exception,[20] it is noticeable that contemporary writers in the Jewish press ignored Tyndall's materialism, which had shocked their Christian counterparts. More generally, Victorian Jewish writers did not criticize science—or, more specifically, the theory of evolution—for encouraging the philosophy of materialism. For these Victorian Jews, the material world was God's creation, as specified in Torah, and there was no compelling reason to consider the world as anything other than material. Rarely was the soul or its immortality discussed in the Jewish press.[21]

Another indicator of Anglo-Jewry's lack of reflection on the soul is the brevity of Michael Friedländer's discussion of the subject in his *The Jewish Religion* (1891). As the long-serving principal of Jews' College, Friedländer was one of the most prominent Jewish educators of his generation, and his

17. Frank Turner, *Contesting Cultural Authority: Essays in Victorian Intellectual Life* (Cambridge: Cambridge University Press, 1993), 196; Bernard Lightman, "Scientists as Materialists in the Periodical Press: Tyndall's Belfast Address," in *Science Serialized: Representations of the Sciences in Nineteenth-Century Periodicals*, ed. Geoffrey Cantor and Sally Shuttleworth (Cambridge: MIT Press, 2004), 199–237.

18. *Dublin Review*, cited in Lightman, "Scientists as Materialists," 210.

19. "Science and Speculation," *JC*, 28 August 1874, 348–49.

20. "Tyndalism," *JC*, 11 December 1874, 589. The author, who may not have been Jewish, makes brief reference to Tyndall's materialism.

21. Marc Swetlitz has pointed out that many American Jews of this period were deeply concerned about materialism, and a number of writers in the American Jewish press criticized Tyndall for his advocacy of materialism: "Responses of American Reform Rabbis to Evolutionary Theory, 1864–1888," in *The Interaction of Scientific and Jewish Cultures in Modern Times*, ed. Yakov Rabkin and Ira Robinson (Lewiston, NY: Mellen, 1995), 103–25.

book probably better reflects educated opinion within Anglo-Jewry than does any other work. This book, he claimed, provided an explication of traditional religious principles. While the second half of this five-hundred-page work was devoted to the duties of the observant Jew, the first half contained Friedländer's commentary on Maimonides' thirteen principles of faith. The thirteenth—"The belief in the revival of the dead, or the immortality of the soul"—was the only place where he addressed the existence of the soul. Although Friedländer cited Saadia Gaon, Maimonides, and Judah Halevi in support of the soul's immortality, he pointed out that this doctrine was not to be found in the Torah. Indeed, he seemed rather uncomfortable in discussing the topic, which he passed over quickly, admitting that we can form only a hazy notion of the resurrection of the dead.[22] In contrast to Friedländer's cursory examination, the soul and its immortality underpinned the doctrines of salvation and the afterlife that were central for nineteenth-century Christians, especially evangelicals. Contemporary Christian periodicals overflowed with references to the soul. Materialism was seen as heretical and a profound threat to Christianity, whereas Anglo-Jewish writers appear not to have been disturbed by a materialistic view of the world. Indeed, while Friedländer insisted on the existence of God the Creator, he identified the main theological threats as arising from polytheism, pantheism, atheism, and deism—not materialism per se.[23]

The second topic concerns design arguments, which featured prominently in Christian works of natural theology, most famously in Paley's *Natural Theology* (1802) and in several of the Bridgewater Treatises. Although evangelicals and High Churchmen tended to accord design arguments less centrality than did other Christians, such arguments were very frequently deployed in nineteenth-century Christian periodicals, in sermons, and in a wide array of scientific books. By contrast, Jewish writers of the period very rarely utilized arguments from design. For example, in the first two decades of publication the *Jewish Chronicle* (f. 1841) contained only one article that dwelled on design. This anonymous contributor reflected on the "great book of nature" and urged parents to inculcate into their children an appreciation of God the Creator by drawing their attention to the design manifested in nature. Moreover, I know of only one book by a Jewish author from the 1840s and '50s that employed design arguments; this was Miriam Mendes

22. Michael Friedländer, *The Jewish Religion* (London: Kegan Paul, Trench, Trübner: 1891), pp. v, 164–67, and 321–22. Like Friedländer's *Textbook of the Jewish Religion* (London: Kegan Paul, 1890), this work passed through many editions.
23. Friedländer, *The Jewish Religion*, 25–29.

Belisario's *Sabbath Evenings at Home* (1856), which was likewise directed at children.[24]

The very noticeable contrast between the prevalence of design arguments in writings by Christians and their absence in works by English Jews invites analysis. There are several possible causes. First, for Jews the Torah and other sacred books take precedence over any other theological source. Thus nineteenth-century Jews, like many evangelical Christians, would have felt no need to deploy design arguments in order to justify the existence and attributes of God. Moreover, such biblical passages as "The heavens declare the glory of God, and the firmament sheweth his handywork" (Psalm 19:2) assert that the world is designed, and therefore Jews need not look to nature for signs of design.[25] Second, while British Christians frequently appealed to the two-book analogy—which views nature as God's other "book" and thereby legitimates design arguments—this analogy and the associated way of thinking about nature were almost totally absent from Jewish works.[26] Finally, the Anglo-Jewish community was principally urban and had no tradition of nature study; hence appeals to natural phenomena were unlikely to prove attractive to most readers whose experience of natural history and of other observational sciences was severely limited.[27]

Although Christians continued to use design arguments after the *Origin of Species* was published, their apologetic functions were severely compromised. As John Hedley Brooke has noted, after the *Origin*, the "image of God as artisan or mechanic . . . took a beating."[28] The *Origin* thus posed a severe challenge to those Christians, principally non-evangelicals, who had

24. A., "On the History of Nature as Conducive to Religion," *JC*, 10 December 1841, 22–23; Miriam Mendes Belisario, *Sabbath Evenings at Home; Or, Familiar Conversations on the Jewish Religion, its Spirit and Observances. Revised Rev. D. A. de Sola* (London: Joel, 1856).

25. Within the Anglo-Jewish context design arguments occurred mainly in educational contexts, as the two examples cited in n. 24 indicate. Two later examples are Ellis Davidson's *The Bible Reader, an Abstract of the Holy Bible* (London: Vallentine, 1877) and Nathan Joseph's *Religion Natural and Revealed* (London: Macmillan, 1879). See Cantor, *Quakers, Jews, and Science*, 308–14.

26. Menachem Fisch, "Reading God's Two Books: Science and the Talmud's Debate on Religion," Selig Brodetsky Memorial Lecture, delivered at the University of Leeds, 30 May 2002.

27. Note the similarity to traditional Jewish attitudes to the environment discussed in Hava Tirosh-Samuelson, ed., *Judaism and Ecology: Created World and Revealed Word* (Cambridge: Harvard University Press, 2002), esp. editor's introduction, p. xxxiv.

28. John Brooke and Geoffrey Cantor, *Reconstructing Nature: The Engagement of Science and Religion* (Edinburgh: Clark, 1998), 161–66. On the many uses of design arguments see Brooke, "The Natural Theology of the Geologists: Some Theological Strata," in *Images of the Earth: Essays in the History of the Environmental Sciences*, ed. Ludmilla Jordanova and Roy Porter (Chalfont St. Giles: British Society for the History of Science, 1997), 53–74.

traditionally placed great emphasis on design arguments. However, as Anglo-Jewish writers had rarely used these arguments, the impact of the *Origin* on this front was minimal.[29]

For the many Christians who considered that the theory of evolution (along with a number of other scientific developments in the mid-Victorian period) attacked their religious beliefs, the issues of materialism and divine design were crucial. As neither topic was prominent within Anglo-Jewry, two of the most controversial aspects of evolution barely impinged on the Jewish community. Therefore, as we shall see in the next section, the Jewish response to evolution was not significantly colored by these two topics, thus enabling British Jews to adopt a more conciliatory attitude towards evolution. This attitude was enhanced by influences arising from the perceived social position of Anglo-Jewry.

The Standard Anglo-Jewish Response to Science

Even before the publication of the *Origin* some English Jews had sought to show the coherence between Torah and science. Although there had been a long tradition among Christian writers of deriving the true system of the world from the King James translation of the Bible, Jewish writers stressed the importance not only of using the original Hebrew text but also of subjecting it to the close analysis that drew on traditional Jewish forms of exegesis. Thus one midcentury Jew complained that Christian theologians who were dependent on the English translation "could not explain the hidden meanings of the Hebrew words." By contrast, the rabbinic commentators knew that "every letter has its own signification and use." Thus only someone who could read Hebrew and had access to the Torah and the rabbinic commentaries could comprehend God's plan as explicated in Genesis, where, as a contributor to the *Jewish Chronicle* put it, "Moses tells us . . . *how* the solar system has been contrived, and who the Contriver is."[30] Jews thus believed that they possessed privileged knowledge of the structure of the world.[31]

29. One who did use such arguments was Nathan Joseph. His *Religion Natural and Revealed* contains an extensive discussion of design and an appendix devoted to "Evolution and Design" (257–59). Joseph, an architect, adopted a rationalist, modernist stance, insisting that the theory of evolution greatly strengthened the argument from design since it demonstrated divine foresight.

30. M. H. S[imonson], *JC*, 3 October 1851, 414.

31. On Anglo-Jewish attitudes to Torah, see David S. Ruderman, *Jewish Enlightenment in an English Key: Anglo-Jewry's Construction of Modern Jewish Thought* (Princeton: Princeton University Press, 2000).

Although few Jews would have embraced the specific cosmological theories propounded by this writer, there was a widely held view within Anglo-Jewry that Jews were uniquely placed to embrace the true convergence between science and religion. For example, in 1875 the author of the "Judaica" column in the *Jewish Chronicle* identified a passage in the Talmud that substantially agreed with Darwin, although it employed specific examples rather than a general theory. This writer directed attention to the talmudic view that God changes the world every seven years, by transforming one animal type into another; for example, "the kunkumah (supposed to designate some species of serpent) [is changed] into a stork, [and] the louse becomes a scorpion." Moreover, in another Talmudic passage humankind was portrayed as originally hermaphroditic but was subsequently "divided into two beings of different sexes." According to this columnist, "a somewhat similar hypothesis was brought forward by Darwin in his 'Origin of Species.' "[32] Although the argument is admittedly weak, it is interesting because this author, like the other Anglo-Jewish writers discussed here, sought to show that Jewish tradition cohered with Darwin's theory.

This strategy of arguing for the coherence between Judaism and science, especially evolution, forms the central plank of what I shall call the standard Anglo-Jewish response, since it was articulated by the majority of Jewish writers from about the mid 1860s until at least the early 1890s. They emphasized that Jews experienced no difficulty encompassing recent developments in science, since science and Judaism are compatible. "Judaism," wrote one commentator, "has nothing to fear from the advancement of Science, but everything to gain."[33] We will shortly examine some of the specific arguments that underpinned this assertion. While some writers focused on evolution, others addressed science more generally, implicitly including evolution as probably the most prominent and controversial aspect of science. A second feature of the standard account was the contrast—often explicitly stated—between the proscientific position of Jews and the immense difficulty that Christians were facing in confronting contemporary scientific developments, especially evolution. Thus the standard response was conditioned by the social pressures discussed in the preceding section. However, as we shall see, this prevalent positive response to science masked deeper uncertainties within Anglo-Jewry.

Some, but by no means all, of the Jewish writers who commented on the convergence between science and Judaism admitted that there were current

32. "Darwin and the Talmud," *JC*, 1 October 1875, 435.
33. "Advancement of Science," *JC*, 14 April 1876, 25.

difficulties but that these would in time be resolved. Perhaps the most interesting example is Friedländer's *The Jewish Religion*, where he discussed several familiar points of opposition between evolution and the biblical account of Creation; for example, whereas the Torah attributes the creation of plant and animal species to specific acts performed by God, according to the theory of evolution creation is a gradual process. Which of these incompatible accounts should we accept? To answer this, Friedländer appealed to history. Whenever Jewish thinkers had encountered secular philosophies, they had neither adopted them uncritically nor rejected them outright in favor of biblical truth. Instead, they had tried to reconcile science with religion. Thus Maimonides had bridged Aristotelianism and Torah, and Moses Mendelssohn had amalgamated Jewish and Enlightenment philosophies. Likewise, Friedländer contended, the theory of evolution would in time be squared with a plausible interpretation of Genesis.[34] Thus, the standard account encompassed both the claim that Judaism coheres with evolution and the belief that any current difficulties will be overcome.

As indicated above, the assurance that science and Judaism were aligned was laced with criticism of Christianity. For example, in an 1864 editorial in the *Jewish Chronicle*, Abraham Benisch (one of the few accomplished Torah and Talmud scholars in Britain) complimented the bishop of London on siding with science and rationality against those contemporary Christians who had adopted an evangelical stance. Benisch insisted that the alliance between Judaism and reason, and therefore science, was sanctioned by such biblical verses as 2 Chronicles 1:10–12, in which God praised Solomon for preferring "wisdom and knowledge" to "riches, wealth, and honour." Moreover, rationality had repeatedly been extolled by the rabbinic tradition.[35] Likewise, in a sermon preached in 1872 Chief Rabbi Nathan Marcus Adler asserted that, in contrast to the apparent antagonism between science and Christianity, "such a contest must be slight or superficial in Judaism, where faith and reason go hand in hand."[36] As Judaism was the natural ally of reason, it could not stifle such products of human reason as Darwin's theory of evolution.

A further example dates from September 1875, when the *Jewish Chronicle* carried an article reacting to a recent event in the Christian world. A prize for the study of theology had been established at a school in Somerset. However,

34. Friedländer, *Jewish Religion*, 30–39.

35. "The Bishop of London on Science and Revelation," *JC*, 18 November 1864, 4–5; 2 December 1864, 4–5.

36. "Science and Religion: A Sermon Delivered by the Rev. Chief Rabbi at the Central Synagogue, on the Sabbath Preceding Pentecost," *JC*, 21 November 1872, 167. See also "The British Association at Norwich," *JC*, 4 September 1868, 4.

the founder had explicitly excluded science from the scope of this prize in the belief that science was inimical to theology. As the writer in the *Jewish Chronicle* pointed out, the founder had indeed been correct to adopt this stance since science is incompatible with Christianity: the former is rational but the latter irrational. Particularly flawed was the Christian doctrine of salvation "according to which millions of human beings are being consigned to ever-lasting perdition" because they are either ignorant of Christianity or have consciously rejected it. On this account God is required to pour out his wrath on the non-Christian, which seems utterly incompatible with his role as "the fountain of justice." As Christians are committed to this utterly irrational view, they must necessarily reject science as incompatible with Christianity. By contrast, added our author, "[t]here is only one theology in existence which is not antagonistic to science—this is Jewish theology." Unlike the New Testament, with its emphasis on the spirit, Judaism had always extolled rationality and the acquisition of knowledge.[37]

A more aggressive tone was adopted in a report on Tyndall's Belfast address by Myer Davis, the editor of the *Jewish World* (which, selling at one penny, sought to undercut the *Jewish Chronicle* by appealing to a broader audience). He suggested that in the light of the prevailing conflict between Christianity and science, the Christian press should "be filled with . . . reports that Christianity is losing its hold upon its professors, and that it is no more than it was of yore." Faced by "the steady advances of Materialism and Free Thought on the one hand, and by Ritualism on the other," Christianity was under attack and Christians were therefore no longer in a position to cast aspersions on the Jews. Erring on the side of caution, Davis advised his readers not to enter the fray between science and Christianity but to stand above the mêlée. Yet, in the light of widespread anti-Jewish and conversionist propaganda, Myers was clearly hoping that the tormentors of the Jewish community would be humiliated by losing their battle with Tyndall, Huxley, and the other scientific naturalists. Judaism, so often vilified by Christians, would emerge the more resilient religion.[38]

Thus the leaders of the well-established and well-assimilated Jewish community, whose voices dominated synagogue pulpits and the Jewish press, did not view evolution as a major threat to Judaism but saw the unassailable alliance between Judaism and science as providing support for Judaism and a timely response to its Christian detractors. They were equally sure that

37. "Science and Theology," *JC*, 24 September 1875, 412. See also "The British Association at Norwich."

38. *JW*, 30 October 1874, 4–5.

Torah would be undiminished by science and that Jews would continue to accept it as the cornerstone of their religion.[39]

Challenges to the Standard Response

The rather complacent assessment of science conveyed by the purveyors of the standard account masked deeper worries about the future of Judaism in England. This is evident in the concern among some communal leaders that their congregants, especially younger members, might be seduced by modern trends in thought, such as evolution, and forsake religious tradition.[40] A similar but more specific concern was expressed in 1868 in the *Jewish Record*. While the *Jewish Chronicle* tended to speak for the Anglo-Jewish establishment and rising middle-class Jews, the *Jewish Record* reflected a more traditional view of Judaism held principally by less assimilated and generally poorer members of the community. In a series of articles on the Pentateuch, one such contributor presented a view of science that challenged the standard Anglo-Jewish response. He pointed out that God had created our minds and our imagination so that we can progress in our understanding of his creation: "these struggles of mind against mind, theory against theory, those aspirations for discovery, that thirst after the unknown, which constitutes intellect—the grandest work of Creation—the true image of its Creator." Yet this writer also considered that the progress of knowledge would lead to moral decay. Although Torah does not provide a true and scientific understanding of creation, he argued, it provides a reasonable account that is "sufficiently likely to be accepted as true by faith that trusts, and by love that adores." Thus Torah is indispensable and must not be undermined by science.[41]

The views of Michael Henry, who edited the *Jewish Chronicle* during the crucial years 1869 to 1875, deserve our attention since, unlike most members of the Anglo-Jewish community, he was well informed about recent developments in science. He had worked as a patent agent, joined the Society of Arts, assisted the editor of the *Mechanics' Magazine*, contributed to the *Mining Journal*, and edited the *Inventors' Almanack*. He was also elected to the Institution of Civil Engineers and attended several annual meetings of the British Association for the Advancement of Science. His professional

39. A similar triumphalism was displayed by some American rabbis. See Swetlitz, "Responses of American Reform Rabbis," 107.

40. "Annual Meeting of the British Association: The President's Address," *JC*, 16 August 1872, 279–80; "Religion and Science," *JC*, 23 August 1872, 290–91; "Science and Speculation," *JC*, 28 August 1874, 348–49.

41. "Pentateuch," *JR*, 25 September 1868, 4–5; 2 October 1868, 5; 27 November 1868, 2.

activities brought him into regular contact with scientists, inventors, and engineers. His familiarity with current scientific news led to a significant increase in the number of scientific books reviewed while he was editor of the *Jewish Chronicle*. Although he was more concerned than were his predecessors with bringing science to the notice of the Jewish community, he was highly respected for his Jewish learning and commitment to traditional modes of Judaism.

Straddling the world of Anglo-Jewry and the communities of scientists and engineers, he, more than other Jews, recognized that scientific naturalists such as Huxley and Tyndall posed a threat not only to Christianity but also to Judaism. In order to understand his response to evolution we need to appreciate his views about the difference between science and technology. In contrast to the speculative nature of scientific theories, he considered that technology rested on a firm foundation of facts and its results could be justified by practice—a view doubtless founded on his experience as a patent agent and editor of technical periodicals. While he supported the introduction of technical subjects into Jews' College School, he remained highly skeptical about the theories used by scientists and wrote critical articles on Darwinism (1871) and on John Tyndall's Belfast address.

In the opening paragraph of his editorial on Darwinism, Henry claimed that Darwin's theory should neither be rejected as "wicked and silly," nor should it be advocated uncritically. Moreover, while it "is opposed to the Biblical account of the creation, . . . it is by no means subversive of the doctrine of the existence of a Creator; nor does it necessitate the supposition that that CREATOR is one whit less powerful than the most pious of us would declare." Yet as the editorial progressed, Henry expressed increasing doubts about the theory, which had not been proved and was incapable of proof. Indeed, it was easily refuted by many arguments. For example, he cited the case of the gorilla—allegedly our closest neighbor biologically—which is significantly different from humans in terms of brain size and the functions performed by certain internal organs.[42] Nor could he find any evidence that humans were more intelligent than they had been in biblical times—an implication, he believed, of the theory of evolution. These counterarguments also indicated the unsettled nature of scientific theories. On his view the current favorite theory would inevitably be rejected at a later date. Henry concluded this editorial by contrasting the futility of scientific theorizing with "the beautiful accuracy of the Biblical account of the creation as proved [*sic*]

42. Henry may have been drawing on arguments developed by the anti-Darwinian comparative anatomist Richard Owen.

by modern scientific research." Thus, despite his initial assertions, he ended by minimizing the importance of Darwin's theory when compared with the robust Genesis narrative.[43]

Henry also became increasingly concerned about the authority that scientists were claiming for themselves and their views. For example, in 1874 he chastised Tyndall for portraying himself as "the priest of a new religion"—the religion of science. Indeed, Henry was clearly concerned lest Tyndall's recent presidential address at the British Association meeting in Belfast would provide scientists with a precedent justifying the exercise of their authority outside the legitimate domain of science. Since in his view science only provided capricious, partial, and uncertain knowledge, he claimed that Tyndall had misled the public by portraying scientific theories as immutable truths whose meaning extended far beyond the facts. In articulating this false view of science Tyndall had set knowledge, "which should ever be the handmaiden of Faith," not only above religion but also in conflict with it. Henry was shocked by Tyndall's chutzpah, reiterating his belief that when science was properly understood, it could not come into conflict with revelation. Indeed, science should "testify to the Greatness of the Creator palpable in the marvels of His workmanship."[44]

Although Henry was generally supportive of science and believed that Judaism did not fundamentally clash with evolution, he was alarmed not only by the excessive claims of the scientific naturalists but also by the utterances of one particularly vociferous member of the Anglo-Jewish community. The dissident was Alfred Gutteres Henriques, a barrister and respected communal leader who served on the Board of Deputies and the Anglo-Jewish Association and later became a deputy lieutenant for the City of London.[45] In order to appreciate his intellectual stance we should note that he belonged to a Reform synagogue—the West London Synagogue. The Reform movement in England was created in the early 1840s when a number of Jews from elite families broke away from the existing synagogues and founded their own synagogue closer to their homes on the west side of London. Although some innovations were introduced at the West London Synagogue, the form of service differed little from mainstream Orthodoxy, as exemplified by the United Synagogue founded in 1870 and headed by the chief rabbi.[46] This situation contrasted sharply with the state of affairs in

43. "Darwinism," *JC*, 15 December 1871, 8–9.

44. "Science and Speculation," *JC*, 28 August 1874, 348.

45. "Alfred G. Henriques," *JC*, 7 August 1908, 7.

46. Anne J. Kershen and Jonathan A. Romain, *Tradition and Change: A History of Reform Judaism in Britain 1840–1995* (London: Vallentine Mitchell, 1995); Alderman, *Modern British Jewry*, 35–37.

Germany and America, where a radical, rationalist form of Reform Judaism had developed in opposition to traditional modes of Judaism. For example, German and American Reform movements generally dispensed both with kashrut—eating kosher food, which was considered outmoded and irrational—and with most, if not all, of the Hebrew liturgy. Prior to the 1890s, when Claude Montefiore, among others, introduced antitraditionalist practices into Anglo-Jewry, British Jews had generally ignored the radical reformist ideas that had taken root in Germany and America. There were a few exceptions, the most notable being Henriques, who was a rationalist, an anti-traditionalist, an evolutionist, and a proponent of the higher biblical criticism. Indeed, Israel Finestein has claimed that Henriques was the "most vocal . . . advocate" of progressive Judaism in Britain during the mid-Victorian period.[47] Such a "progressive" thinker could not encompass the standard Anglo-Jewish response to evolution, which diplomatically ignored the challenge that progressive modes of thought posed to traditional Judaism.

In 1869 Henriques wrote to the *Jewish Chronicle*, arguing forcefully that Judaism is totally incompatible with the "Law of Evolution." For example, he claimed that in contrast to evolution, which posits continual change and mutability, "Jewish teaching insists on the very opposite mode of thought. Judaism is based on immutability." He also argued that traditional forms of Judaism were outmoded in several other respects. Scientific rationalism denied the legitimacy of a hereditary priesthood, repudiated the biblical requirement of public sacrifices, and opposed the traditional legal and economic systems sanctioned by Jewish tradition. Indeed, although the central dogma of God's unity still remained intact, modern science undermined so many facets of traditional Judaism that Henriques doubted whether it could ever be revivified as a religion after Darwin's onslaught.[48]

The editor, Michael Henry, was clearly so distressed by this forthright letter that he took the unprecedented step of prefacing it with remarks intended to refute Henriques's arguments. To bolster his initial remarks he added a more extensive refutation in the following week's number. Contrary to Henriques's alignment of science with truth, Henry (again) presented science as "feeble," since its theories were not immutable; history showed that a theory accepted at one time will be rejected at a later date and replaced by another. In contrast to the vicissitudes of science, proclaimed Henry,

47. Israel Finestein, *Anglo-Jewry in Changing Times: Studies in Diversity 1840–1914* (London: Vallentine Mitchell, 1999), 99.

48. Alfred Gutteres Henriques to the editor, *JC*, 20 August 1869, 8.

"Judaism is eternal": it has "withstood the scientific enquiries of centuries" and would not succumb to any present-day opponent. He then asserted, perhaps inconsistently, that "every advance made by science demonstrates the truth and vitality of Mosaic Revelation," and, reiterating the standard response, insisted that science and Judaism are fully convergent.[49]

These remarkable editorial interventions indicate that Henriques was seen to pose a significant threat to the community and to Jewish tradition. He had rejected the much-vaunted harmony between science and Judaism that communal leaders had both sought to foster and consistently deployed in constructing their opposition to both anti-Semitism and the conversionists. Most importantly, he had challenged the liberal ideal that a Jew could be an acculturated English gentleman immersed in secular learning. The implication of Henriques's argument was that Jews had to choose between an outmoded religion, based on tradition, and the modern world, which included the theory of evolution.

In 1891 Henriques again generated a controversy in the *Jewish Chronicle* that illuminates growing tensions within Anglo-Jewry during the closing years of the century. In an editorial Asher Myers had warmly praised a widely discussed recent article by the duke of Argyll, who was not only a supporter of several Jewish causes but also a prominent antievolutionist.[50] Argyll was applauded for offering "scientific proof as to the possible truth of the Biblical narrative of the Deluge" and for condemning "Professor Huxley's derisory denial of it [which] is shown to be of no scientific value whatever." The *Chronicle*'s support for Argyll brought a sharp response from Henriques, who contrasted the learned Huxley, whose "splendid reputation" stands "second to none in this country," with Argyll's "feeble attempts" to support Scripture. "I thought," added Henriques, that "the Noachian Deluge had been disestablished long since through the labours of our geologists and others." He then repeated some of the familiar arguments that Huxley had used against the biblical account of the Deluge; for example, there is no adequate explanation of where the water came from or how it was subsequently dispersed.[51]

Myers's support for Argyll is surprising in the light of the *Jewish Chronicle*'s generally proscience and pro-evolution stance, as argued above. As Argyll had strongly supported Jewish emancipation, Myers may have been prompted by political expediency. He may also have wanted to appeal to the

49. "The Scripture and Science," *JC*, 27 August 1869, 5. A far more moderate critique of Henriques appeared on p. 3 in a letter by "Ryall."

50. [George Campbell, duke of] Argyll, "Huxley on the Warpath," *Nineteenth Century* 29 (1891): 1–33.

51. *JC*, 6 February 1891, 6; Alfred Gutteres Henriques to the editor, 13 February 1891, 8.

large number of devout Eastern European Jews who had arrived during the previous decade. Whatever the reason, Henriques was clearly surprised by the editor's naïveté in supporting a prominent antievolutionist and Christian apologist. By contrast, Henriques considered that the Jewish community should turn its back on a tradition that could not be justified by reason and should instead encompass modernity. Only a thoroughly reformed and rational form of Judaism could survive in the modern age. Not surprisingly, Henriques encouraged Claude Montefiore in his attempts to import progressive Judaism into Britain.[52]

Raphael Meldola: The Jewish Darwinian[53]

We return to examine the writings of Raphael Meldola, who published principally on entomology and was the foremost established Jewish scientist of this period who strongly supported Darwin's theory. He was also a committed Jew who remained throughout his life a subscribing member of the Spanish and Portuguese Synagogue in London, where his grandfather had served as ḥakham (communal leader). One of the earliest members of the Maccabæans, he served for a time as its president. Despite his clear commitment to Judaism he was not a very observant Jew. For example, he undertook field trips on the Sabbath.[54] Also, in a letter of 1886 to Sir Joseph Sebag-Montefiore, president of the Spanish and Portuguese Synagogue, Meldola indicated that his commitment to science had "precluded the possibility of my taking an active part in the affairs of the synagogue."[55]

We should also consider Meldola's position within the familial structure of Anglo-Jewry. His father's family were Sephardim, and although Sephardim formed a small minority of declining influence during the Victorian period, they had previously constituted the upper echelon of Anglo-Jewry. They were generally well-traveled and intellectually outward looking. For example, as David Ruderman has argued, David Nieto, the ḥakham during the early eighteenth century, sought to bridge tradition and modern learn-

52. Lucy Cohen, *Some Recollections of Claude Goldsmid Montefiore 1858–1938* (London: Faber and Faber, 1940); Steven Bayme, "Claude Montefiore, Lily Montagu, and the Origins of the Jewish Religious Union," *Transactions of the Jewish Historical Society of England* 27 (1980): 61–71.

53. The only extended biography is James Marchant, ed., *Raphael Meldola* (London: Williams and Norgate, 1916). See Cantor, *Quakers, Jews, and Science*, 340–45.

54. A. R. Wallace to Raphael Meldola, 19 June 1883, Hope Collection, Oxford University Museum of Natural History (hereafter OUM).

55. Raphael Meldola to Joseph Sebag-Montefiore, 2 December 1886, in Marchant, *Raphael Meldola*, 148–49.

ing.[56] Likewise, a century later Meldola's grandfather, Ḥakham Meldola, was a friend of the astronomer William Herschel and was no "stranger to the cultivation of science, which, as an enlightened man, he pursued with considerable success."[57] Meldola may thus be interpreted as continuing this tradition among the Enlightened Sephardim. However, it was his mother, Matilda (Teltsel), who particularly encouraged his scientific interests and accompanied him on his early entomological field trips. In a letter written shortly after her death in 1892 he reflected: the "history of my early life is one of the most remarkable cases of the indebtedness of son to mother for everything that I have been able to accomplish in life."[58] Matilda's scientific and intellectual interests—she played chess with Alfred Russel Wallace when he visited the Meldola household[59]—may have been inherited from her Ashkenazi father, Moses Abraham, who was an optician and thus one of the few English Jews with a science-related occupation.

"[A]s a youth, I fell into the ranks of Darwinians," Meldola later reminisced.[60] The surviving correspondence with Charles Darwin, numbering seventy-five letters during the period 1871 to 1882, indicates that Meldola adopted Darwin as his scientific mentor.[61] Darwin also aided Meldola's career by heading the list of proposers for his fellowship of the Royal Society.[62] Meldola likewise developed close friendships with Alfred Russel Wallace and several other leading "Darwinians," including Edward Poulton, who was appointed Hope Professor of Zoology at Oxford and shared Meldola's antipathy to Lamarckian and mutationist accounts of evolution. Thus Meldola was not just an evolutionist; he firmly aligned himself with the Darwin circle and, after Darwin's death in 1882, with the neo-Darwinians who remained faithful to Darwin's natural selectionist program.[63]

56. David S. Ruderman, *Jewish Thought and Scientific Discovery in Early Modern Europe* (New Haven: Yale University Press, 1995), 310–31.

57. *HO* 1(1853): 115.

58. Raphael Meldola to Edward Poulton, 4 February 1892, Hope Collection, OUM.

59. A. R. Wallace to Raphael Meldola, 14 May 1886, Wallace Correspondence, OUM.

60. Raphael Meldola, "The Presidential Address; Delivered . . . at the Annual Meeting, January 27th, 1883," *Transactions of the Essex Field Club* 3 (1884): 59–93.

61. Frederick Burkhardt, ed., *A Calendar of the Correspondence of Charles Darwin, 1821–1882* (Cambridge: Cambridge University Press, 1994).

62. Certificate EC/1886/10, Archive of the Royal Society of London.

63. Cf. St. George Jackson Mivart, a Catholic, who questioned the adequacy of natural selection and became embroiled in controversy with some leading Darwinians. Jacob W. Gruber, *A Conscience in Conflict: The Life of St. George Jackson Mivart* (New York: Columbia University Press, 1960).

Under Darwin's guidance Meldola developed his research on mimicry, particularly the study of butterflies that possessed the coloring and appearance of a more aggressive local species. He recognized that the theory of evolution offered a ready explanation of this phenomenon since, in adopting the physical characteristics of another species that was less subject to predators, the insect's survival could be enhanced. It is clear that by 1871, if not earlier, Meldola had been attracted to this problem precisely because it required the application of natural selection and was, in turn, an impressive confirmation of Darwin's theory.[64] Several of Meldola's subsequent publications developed from his correspondence with Darwin. For example, after informing Darwin that he was researching the problem of mimicry, Darwin forwarded him a letter recently received from Fritz Müller, a German naturalist working in Brazil. So impressed was Meldola with Müller's application of Darwinian principles that he cited lengthy extracts from his letter in an 1878 paper entitled "Entomological Notes Bearing on Evolution," in which he drew attention to several remarkable phenomena that were explicable on evolutionary principles. For example, he cited the case of an Indian mantis— *Gongylus gongylodes*—that simulates the appearance of a flower. This example was, he asserted, of the "highest interest" and it "can, in fact, be only completely appreciated by the believer in *natural selection*."[65]

One of the subjects discussed during the correspondence with Darwin was seasonal diamorphism, where two forms of the same species exist, exhibiting different characteristics at different seasons. For an impressive explanation of the seasonal diamorphism of butterflies Darwin directed Meldola to read August Weismann's *Studien zur Descendenz-Theorie* (1875). Stimulated by Weismann's book Meldola proposed undertaking a translation, which Darwin strongly encouraged. When his annotated English edition was published in 1882, Meldola's own introduction was sandwiched between a preface by Darwin and a second preface that Weismann had written specifically for this English edition.[66]

In addition to this translation, Meldola not only published a number of papers that made incisive use of evolution but he also forcefully expressed his

64. Raphael Meldola, "Mimicry in the Insect World," *Land and Water* 11 (1871): 321–22; Meldola, "Mimicry between Butterflies of Protected Genera," *Annals and Magazine of Natural History*, ser. 5, 10 (1882): 417–25. See also Meldola, "On a Certain Class of Cases of Variable Protective Colouring in Insects," *Proceedings of the Zoological Society* 2 (1873): 153–62.

65. Raphael Meldola, "Entomological Notes Bearing on Evolution," *Annals and Magazine of Natural History*, ser. 5, 1 (1878): 155–61. Emphasis added.

66. August Weismann, *Studies in the Theory of Descent*, trans. Raphael Meldola, 2 vols. (London: Sampson, Low, Marston, Searle and Rivington, 1882).

Darwinian commitments in his contributions to *Nature.* For example, he dismissed as utterly misguided an antievolutionary tract published by the Catholic Truth Society: "So far as the author's attacks are directed against evolution as a principle, his weapon is as a bladder of air against the hide of a hippopotamus."[67] But he reserved his most vigorous criticisms for those who courted forms of evolutionary theory that did not deploy the classic mechanism of natural selection. Thus he criticized George John Romanes for attacking natural selection and for proposing instead his own theory of "physiological selection." Writing in *Nature* Meldola argued that Romanes had misunderstood natural selection and that his alternative theory was scientifically and methodologically flawed.[68] As he confided to a fellow Darwinian, by propounding his theory of "*fizziological selection*" Romanes "has done more harm to the Darwinian theory than any other writer."[69]

Meldola considered that the theory of evolution by natural selection was in full accord with the highest standard of scientific method. Darwin, he claimed, had been the first to subject the species question to the scientific method that had proved so successful in the physical sciences.[70] (We should remember that Meldola was also an astronomer, having worked under Norman Lockyer at the Solar Physics Observatory in the mid-1870s, and a chemist.) The theory of evolution by natural selection required only the minimum of assumptions, all of which could be justified independently of the theory. It therefore offered an uncompromisingly naturalistic and materialistic account of species change. He often referred to it as the "theory of adaptation" and conceived of its operating slowly and inexorably. Moreover, it operated not only on biological species but also "upon internal organisation; minute constitutional or physiological deviations at present utterly beyond the ken of science, can be seized upon and perpetuated by this agency when of any advantage to the possessor. The survival of the fittest is utilitarianism *in excelsis.*" One of the theory's main strengths was its applicability to

67. Raphael Meldola, "On a Certain Class"; Meldola, "Evolution: Old and New," *Nature* 80 (1909): 481–85; Meldola, "Lamarckism *versus* Darwinism," *Nature* 44 (1891): 441–43.

68. Raphael Meldola, "Physiological Selection and the Origin of Species," *Nature* 34 (1886): 384–85. See G. J. Romanes, "Physiological Selection: An Additional Suggestion on the Origin of Species," *Journal of the Linnaean Society* 19 (1886): 337–411; A. R. Wallace, "Romanes versus Darwin: an Episode in the History of the Evolution Theory," *Fortnightly Review* 46 (1886): 300–16. Also John C. Lesch, "The Role of Isolation: George J. Romanes and John T. Gulik," *Isis* 66 (1975): 483–503.

69. Meldola to E. B. Poulton, 17 September 1889 and 15 May 1891, Meldola file, Hope Collection, OUM.

70. Meldola, "Evolution," 481.

many areas of science where it subsumed previously unconnected facts to a single conceptual scheme.[71]

While Meldola was convinced that natural selection was the primary means of species change in nature, he was prepared to acknowledge that other mechanisms might also operate, but he insisted that they played a less significant role. Natural selection was to be clearly distinguished from these other theories, in particular mutationism and Lamarckism. He criticized mutationist accounts of evolution, which were gaining popularity in the early years of the twentieth century, on several grounds: for example, in explaining large and sudden variations in a species mutationists evoked mysterious causes that were not subject to physical laws. Lamarckianism was also criticized not only because it was unable to explain many phenomena, such as the variable protective coloring of insects, but also because some of its proponents appealed to vaguely formulated environmental and even mental factors, thereby failing to follow the proper scientific method.[72] As he wrote to a fellow entomologist: "*Of course* there is not a vestige of Lamarckism in my notion" of seasonal diamorphism.[73] By the time of the 1909 Darwin centenary celebrations he described himself, in the epigraph that heads this paper, as "too 'Darwinian' " and out of kilter with the younger generation of biologists.[74]

I have argued that Meldola adopted a robust interpretation of evolution: one that was often associated with materialism, irreligion, and atheism. Yet he retained his identity as a Jew. It is difficult to know how he conceived the relationship between his science and his religion since he did not write specifically on science and religion. However, with the aid of a few relevant passages we can shed a little light on this topic.

As indicated above, Meldola appreciated the profundity and power of scientific analysis and claimed that in its secular, scientific approach the *Origin of Species* had been "the first work which successfully rescued the species question from the domain of ancient mysticism and ecclesiasticism."[75] This implies that the successful practice of science required its complete separa-

71. Meldola, "Presidential Address," 81.

72. Raphael Meldola, "Lamarckism *versus* Darwinism," *Nature* 38 (1888): 388–89. British neo-Lamarckians included St. George Mivart (*Essays and Criticisms* (London: Osgood, McIlvaine, 1892)) and the Quaker Abraham Bennet ("The Theory of Natural Selection from a Mathematical Point of View," *Nature* 3 (1870–71): 30–33).

73. Meldola to Frederick Dixey, 1897, Hope Collection, OUM.

74. Edward Poulton's *Charles Darwin and the Theory of Natural Selection* (London: Cassell, 1909) was dedicated to Meldola and contains some of Darwin's correspondence with Meldola.

75. Meldola, "Evolution," 481.

tion from religion. Meldola was particularly appalled by the superstition and bigotry of those who attacked science in the name of religion. In his 1883 presidential address to the Essex Field Club, Meldola went further by stating that "Darwin has exalted our conception of Nature beyond the theologies. He has taught us that there is no intermediate and direct interference with the course of natural law—he has enforced the lesson that in studying natural science we are concerned only with secondary causes."[76] He appears to have viewed the universe as mechanistic and law-like, and considered science to exclude any form of divine intervention (or, indeed, any sudden arbitrary intervention, as required by the mutationist theory). On this materialist account, physical effects are produced only by physical causes. The evolution of organisms—humans included—is a purely physical process, leaving no room for an imminent God working in the physical world.

But did Meldola accept God as creator of the physical universe? Although he does not address this directly, one extant letter sheds some light on his attitude toward materialism. It was written in 1886 to Edward Poulton, who clearly approved the synthesis between neo-Darwinism and Anglo-Catholicism that was proposed by the Oxford theologian Aubrey Moore.[77] Poulton had been responsible for the Meldolas hiring a servant named Judy who died suddenly. Reporting her death to Poulton, Meldola asked playfully "[w]hether her 'soul had gone aloft' "? To which he added: "I know not as I must leave problems of that kind to the Oxford theologians who are so well endowed that they must, 'by virtue of their office' know all about such things."[78] The skeptical, teasing tone of this letter suggests that Meldola either denied the existence of the human soul or considered discussion of such issues at best unproductive and at worst pointless. Referring back to an earlier section of this paper, we may also take Meldola's comment as an example of the contrast between the theological centrality of the soul for Victorian Christians and the relative lack of interest in the topic among English Jews.

From the admittedly rather limited evidence available, Meldola appears not to have seen any incompatibility between being an evolutionist and materialist and being a Jew, albeit a fairly assimilated one. Can Meldola, then, be understood as aligning himself with those Jews who articulated the standard response, which stressed the coherence between Judaism and science? Only in a very limited sense, since he appears never to have committed

76. Meldola, "Presidential Address," 91–93.

77. Richard England, "Natural Selection, Teleology, and the Logos: From Darwin to the Oxford Neo-Darwinians, 1859–1909," *Osiris* 16 (2001): 270–87.

78. Raphael Meldola to Edward Poulton, 28 December 1886, Hope Collection, OUM.

himself to this position in public. Moreover, unlike the rabbis and Jewish communal leaders who articulated the standard response as a public statement about the place of Jews in Victorian England, he saw himself first and foremost as a practicing scientist and as a member of the scientific elite. As Hannah Gay has noted in discussing Meldola's attempt to place a plaque in Westminster Abbey to commemorate Herbert Spencer, he tended to be "rather exclusive" and sought to network with people in "influential positions in scientific, academic and public life."[79] Although he taught numerous students at Finsbury College and also offered occasional lectures to Jewish workingmen,[80] he firmly aligned himself with the scientific elite, and especially Darwin and the neo-Darwinians. Both socially and intellectually he was a Darwinian.

79. Hannah Gay, "No 'Heathen's Corner' Here: The Failed Campaign to Memorialize Herbert Spencer in Westminster Abbey," *British Journal for the History of Science* 31 (1998): 41–54; Geoffrey Cantor, "Creating the Royal Society's Sylvester Medal," *British Journal for the History of Science* 37 (2004): 75–92.
80. *JC*, 17 December 1875, 611; 12 January 1877, 6.

2

Responses to Evolution by Reform, Conservative, and Reconstructionist Rabbis in Twentieth-Century America

Marc Swetlitz

In 1992, Rabbi Arthur Green, then president of the Reconstructionist Rabbinical College, lamented that Jews were now convinced that the "origin of species—and of the universe itself" is to be "explained by scientists rather than theologians." Reflecting upon Jewish thought in the twentieth century, Green wrote that Jewish theologians had "largely abandoned Creation as a theological issue" and saw "no value" in defending "ancient Jewish views on Creation."[1] While he dismissed literal interpretations of Genesis, Green forcefully maintained that God was involved in the origin of the universe and of species. And he urged Jewish thinkers to once again make the ongoing creation of the cosmos and of life a central theological concern and to explore how science and religion could be made relevant to one another.

While the evolution of life had been ignored by most prominent twentieth-century Jewish theologians, it was an important subject for the broader community of Reform, Conservative, and Reconstructionist rabbis, at least until the end of the 1960s. In sermons, popular articles, books, and presentations at rabbinical conventions, several of America's leading rabbis addressed the relationship of the scientific study of evolution to Judaism. Thus one aim of this paper is to provide a historical narrative for the period between the nineteenth-century American rabbinic engagement with evolution and the present. For that period I have identified eight episodes, from the Scopes trial and the rise of Protestant fundamentalism in the 1920s to a renewed interest in Jewish mysticism and kabbalah in the 1980s and '90s. In each case I will explore the social and political contexts, as well as intellectual developments, both inside and outside the Jewish community in order to understand

I thank Geoffrey Cantor, Arnold Eisen, Ron Numbers, Michael Ruse, and Jonathan Sarna for their invaluable comments and suggestions. In addition, I am indebted to the assistance of librarians at the American Jewish Archives, the Asher Library, the Mordecai Kaplan Library and Archives, and the Naperville Public Library.

1. Arthur Green, *Seek My Face, Speak My Name* (Northvale, NJ: Aronson, 1992), 53.

why rabbis were interested in evolution, how they engaged it, and the broader social and intellectual meanings they attributed to it.

As a survey of key individuals and episodes across the twentieth century, this chapter provides a basis for identifying and exploring common themes and concerns. I will focus on three topics that the rabbis in this study understood to be the primary points of connection between evolution and Judaism—the nature of God, human nature, and the issue of evil. While the rabbis I discuss were not highly trained in science, they inherited from their nineteenth-century predecessors a conviction that science is an important part of contemporary Judaism, in regard to both the critical study of history and texts and the integration of scientific knowledge with Jewish faith. In addition, they inherited a theistic, progressive view of evolution that integrated elements of scientific theory with God's creative power.[2] This inheritance set the stage for discussions about evolution in the 1920s as well as for the challenges that emerged later in the century.

The Scopes Trial, Protestant Fundamentalism, and Reform Rabbis

The legal, cultural, and theological debates generated by William Jennings Bryan and Protestant fundamentalists in the early 1920s, culminating in the 1925 Scopes trial, attracted the serious attention of leading Reform rabbis, both in sermons and as leaders of the Central Conference of American Rabbis (CCAR). These rabbis viewed the controversy over evolution in the context of their broader concern that Protestant fundamentalists intended to crush freedom of thought, stifle scientific inquiry, and undermine liberal political and social values. President Abram Simon proclaimed at the 1925 CCAR annual convention that his agenda was to preserve the "progressive spirit of all departments of human culture." Simon had previously appointed a committee to prepare a resolution on the relationship of science and Judaism, with a focus on the controversial subject of evolution. Yet the discussion at the 1925 convention revealed considerable disagreement over the statement's content: should it include a theology of evolution; an opinion on how Reform Judaism should interpret Genesis; a statement on how Christian

2. Naomi Cohen, "The Challenges of Darwinism and Biblical Criticism to American Judaism," *Modern Judaism* 4 (1984): 121–57; Marc Swetlitz, "Responses of American Reform Rabbis to Evolutionary Theory, 1864–1888," in *The Interaction of Scientific and Jewish Cultures in Modern Times*, ed. Yakov Rabkin and Ira Robinson (Lewiston, NY: Mellen, 1995), 103–25; Swetlitz, "American Jewish Responses to Darwin and Evolutionary Theory, 1860–1890," in *Disseminating Darwinism: The Role of Place, Race, Religion, and Gender*, ed. Ronald L. Numbers and John Stenhouse (Cambridge: Cambridge University Press, 1999), 209–46.

theology generated conflicts with evolution not faced by Judaism; or only the view that teachers should be free to teach evolution in public schools? The rabbis reached no consensus and the CCAR did not adopt any resolution. But interest was raised in the topic, and the Harvard scholar Harry Wolfson was asked to deliver a paper the following year entitled "Historic and Philosophic Attitude of Judaism to Science."[3]

The views on evolution expressed by Reform rabbis in the 1920s were virtually identical to those of Kaufmann Kohler and Emil Hirsch, who helped to shape the American Reform response to evolution in the late nineteenth century.[4] Indeed, some rabbis were outraged that they had to revisit debates they thought had been settled decades earlier. For Reform rabbis, fundamentalists had erred in reading literally the opening verses of Genesis and in mistakenly assuming that evolution denied a Creator. Reform rabbis concurred that Genesis was not a textbook for science and literal interpretations of Torah were not acceptable. Indeed, evolution meant the progressive development of life from simple to complex forms, including the progressive development of mind and spirit. Human beings stood at the pinnacle of creation, being God's partner in the process of continuous creation. While admitting that the order of nature did not unequivocally prove God's existence or attributes, Reform rabbis continued to proclaim that progressive change and design were characteristic of evolution, and that these features provided a strong case for God as Creator and for God's providence.

Natural selection was also incorporated into their understanding of evolution. Rabbi Samuel Goldenson, who was on the CCAR committee created by Simon, captured this integrated vision in a 1922 sermon: "Everywhere there is the same continuous law of change, transformation, complexity,

3. Leo Franklin, "Report of Committee on President's Message," *CCAR Yearbook* 35 (1926): 174–77; Abram Simon, "Message of the President," *CCAR Yearbook* 35 (1926): 220–24, on 221; Samuel Schulman, "Statement Prepared by Rabbi Samuel Schulman," *CCAR Yearbook* 35 (1926): 177–79; Isaac E. Marcuson, "Report of the Recording Secretary," *CCAR Yearbook* 36 (1927): 24–25; Harry A. Wolfson, "Judaism in Relation to Philosophy and Science," *Jewish Exponent* (Philadelphia), 12 July 1926, cited in Israel H. Levinthal, *Judaism: An Analysis and an Interpretation* (New York: Funk and Wagnalls, 1935), 79 n. 12. See also Stephen J. Goldfarb, "American Judaism and the Scopes Trial," *Studies in the American Jewish Experience II*, ed. Jacob R. Marcus and Abraham J. Peck (Lanham, MD: University Press of America, 1981), 33–47.

4. Emil Hirsch, "Evolution," in *The Jewish Encyclopedia*, ed. Isidore Singer, 12 vols. (New York: Funk and Wagnalls, 1901–6), 4:281–82; Kaufmann Kohler, *Jewish Theology Systematically and Historically Considered* (New York: Macmillan, 1918), 71, 100, 147–51, 154, 211, 216; Swetlitz, "Responses of American Reform Rabbis," 109–14.

competition, natural selection and the survival of the fittest."[5] While recognizing that natural selection involved chance, struggle, and competition, these Reform rabbis, like their nineteenth-century predecessors, placed an optimistic interpretation on natural selection—as the mechanism by which progress had occurred. Applying evolutionary theory to Jewish history, some pointed to the survival of the Jewish people, despite numerous challenges, as indicating that evolution could be a source of hope.[6] This contrasts with the views of Bryan and many Protestant fundamentalists, who considered natural selection to be a source of deep concern, both because of its association with materialism and chance (rather than design) and because of its association with German militarism.[7]

Secularism, the Reorientation of Reform Judaism, and the New Physics

In the late 1920s and '30s, Reform rabbis continued to discuss evolution, but now with new concerns about growing secularism, both inside and outside the Jewish community. Some claimed that the synagogue was "being invaded by secularism." Secular Yiddish and Hebrew organizations expanded, while religious education and practice declined. Throughout America, Marxism and socialism expanded their Jewish following.[8] Reform rabbis were also disturbed to find that Clarence Darrow, a staunch defender of progressive social and political values, had taken to the lecture circuit proclaiming that evolution and religion were incompatible. In response, Rabbis Barnet Brickner, Maurice Eisendrath, and Ferdinand Isserman publicly debated with Darrow and argued not only that evolution and Judaism were compatible but that evolution implied belief in God.[9]

5. Samuel Goldenson, "Evolution and Religion" [12 March 1922], 6, American Jewish Archives, Hebrew Union College, Cincinnati (hereafter AJA), Manuscript Collection #81, Box 1, Folder 3.
6. Goldenson, "Evolution and Religion"; Samuel Goldenson, "Evolution and Religion" [1922], *World Problems and Personal Religion* (Pittsburgh: Rodef Shalom Congregation, 1975), 79–85; Ferdinand Isserman, "Evolution and Religion," 25 July 1925, AJA, Manuscript Collection #6, Box 11, Folder 3; David Philipson, "Evolution and Religion" [1925 or 1926], AJA, Manuscript Collection #35, Box 4, Folder 5; Maurice N. Eisendrath, "The Never Failing Stream" [1930], in *The Never Failing Stream* (Toronto: Macmillans in Canada, 1939), 384–98.
7. Ronald L. Numbers, *The Creationists: The Evolution of Scientific Creationism* (New York: Knopf, 1992), 41–42.
8. Jonathan Sarna, *American Judaism* (New Haven: Yale University Press, 2004), 223–27.
9. "Is Man a Machine? A Debate between Clarence Darrow and Rabbi Barnet R. Brickner" [9 February 1928], AJA, Manuscript Collection #98, Box 3, Folder 1; Maurice Eisendrath, "Can Faith Survive?," in *The Thoughts and Afterthoughts of an American Rabbi* (New York: McGraw-Hill, 1964), 292–95; Ferdinand Isserman, "Genesis Misconstrued in 'Inherit the Wind' " [5 October 1956], AJA, Manuscript Collection #6, Box 20, Folder 2.

The 1930s was also a decade of reorientation for Reform Judaism, marked by greater emphasis on ceremonial practice, Zionism, and belief in a more personal God. The 1937 Columbus Platform reflected this shift with a new set of guiding principles for the Reform movement. At the same time, a vocal group of Jewish humanists emerged who rejected a transcendent, omnipotent, and providential God and envisioned "a finite deity battling for humanity from within the natural world."[10] While this led to heated debates among Reform rabbis, all shared both a desire to counter the growth of secularism and a conviction that science could serve that end as an integral part of a renewed theology. Thus Samuel Cohon, chair of Jewish theology at Hebrew Union College and the primary author of the Columbus Platform, insisted not only that science was among the most important sources of truth in the contemporary world, but that theology had to "coordinate" religious truth "with other manifestations of truth," science included.[11]

As part of this effort, Reform rabbis turned to the popular writings of physicists Arthur Eddington, James Jeans, Robert Millikan, and Arthur H. Compton. While there was no consensus over the philosophical implications of modern physics, these physicists all believed that quantum theory and relativity overthrew the nineteenth-century materialist view of nature. The concept of energy, the interactions between observer and observed in quantum-level measurements, and the highly mathematical character of physical theory were interpreted as pointing towards an idealist philosophy of nature. Most importantly, these physicists interpreted the new physics as supporting a spiritual foundation for reality and a view of God as an immanent force in the evolution of life.[12] In the early 1930s, quotations from these scientists appeared in the writings of Rabbis Cohon, Barnet Brickner (spokesperson for the religious humanists), and Felix Levy (a leading intellectual in the Reform movement). They conceived the new physics as supporting the view of an immanent intelligence operating within nature and moving the evolution of life towards greater consciousness.[13] Moreover, this account of how

10. Michael A. Meyer, *Response to Modernity* (Oxford: Oxford University Press, 1988), 317; Sarna, *American Judaism*, 249–55.

11. Samuel S. Cohon, *Jewish Theology: A Historical and Systematic Interpretation of Judaism and Its Foundations* (Assen, the Netherlands: Van Gorcum, 1971), 18.

12. Peter J. Bowler, *Reconciling Science and Religion: The Debate in Early Twentieth-Century Britain* (Chicago: University of Chicago Press, 2001), 101–4, 255.

13. Barnet R. Brickner, "The God-Idea in the Light of Modern Thought and Its Pedagogic Implications," *CCAR Yearbook* 40 (1930): 304–22; Cohon, *What We Jews Believe* (Cincinnati: UAHC, 1931), 142–52; Cohon, "Has Modern Science Banished God?" [November 1934], AJA, Manuscript Collection #276, Box 19, Folder 2; Felix A. Levy, "The Nature and Scope of Jewish Theology and Its Bearing on Modern Thought," *CCAR Yearbook* 41 (1931): 338–54.

the divine operated in evolution was strikingly similar to the earlier discussions of evolution by Kohler and Hirsch, illustrating the persistence of this view even as Reform theology developed in other areas.

Some Reform rabbis also looked to evolution to support their views about human nature. In *What We Jews Believe* (1931), Cohon argued that Darwin's theory explained the aggressive, destructive, and selfish aspects of human nature, and he pointed to the carnage of World War I as a "poignant example of the brutal tendency in man's nature." At the same time evolutionary theory made it clear that the human attributes of mercy and love have their source in the "tenderness" and "mother instinct" of animal life.[14] In addition, the study of human evolution showed that humans are unique, a view Cohon supported by evoking Thomas Huxley's 1863 essay "Man's Place in Nature." Huxley had argued for the continuity between humans and primates, but he also proclaimed that the evolution of intelligence and moral conscience created a vast "gulf between civilized man and the brutes."[15] Cohon agreed, but he believed this scientific account of evolution was insufficient to understand human uniqueness. The key moment in human evolution was the "dawn of religious consciousness," when humans began to regard themselves as "no longer mere animal but endowed with a divine spirit." Indeed, this recognition was possible only because of the evolution of the human soul, which had its source in God. For Cohon, this insight into human evolution and uniqueness came from Judaism, not science.[16]

Conservative Judaism and the Challenge of Kaplan's Reconstructionism

Conservative rabbis in the 1920s and '30s engaged evolution less frequently than did their Reform contemporaries. They focused on topics related to Jewish law, while virtually ignoring general theological issues.[17] And since reference to evolution typically appeared in discussions about God or human nature, absence of theological discussion helps explain the lack of attention to the topic of evolution. The annual meetings of the Rabbinical Assembly, the association of Conservative rabbis, contained no discussions of evolu-

14. Cohon, *What We Jews Believe*, 159, 174.

15. Ibid., 174–75. On Huxley, see Sherrie L. Lyons, *Thomas Henry Huxley: The Evolution of a Scientist* (Amherst, NY: Prometheus, 1999), 209–13.

16. Cohon, *What We Jews Believe*, 178.

17. Robert G. Goldy, *The Emergence of Jewish Theology in America* (Bloomington: Indiana University Press, 1990), 7–9. This tendency continued to some extent after World War II. See Bernard Martin, "Conservative Judaism and Reconstructionism," in *Movements and Issues in American Judaism*, ed. Bernard Martin (Westport, CT: Greenwood Press, 1978), 132–33.

tion, and only a few references to evolution appeared in published sermons and articles over this period. In two Friday evening lectures (1932 and 1933), Israel Levinthal, rabbi at the Brooklyn Jewish Center and former president of the Rabbinical Assembly, argued that the scientific understanding of evolution reveals order, design, and intelligence in nature. Levinthal made reference to Cohon's *What We Jews Believe* and quoted from prominent physicists, including Millikan, Jeans, Eddington, and Einstein.[18] Later in the decade, in 1938, the physicist Arthur Compton was invited by Louis Finkelstein to lecture at the Jewish Theological Seminary (JTS), where Finkelstein was provost. Compton explained how physics and evolutionary biology formed the basis for belief in God as an immanent power underlying the progress of evolution, "culminating in man's spiritual nature and God-like powers." Finkelstein's theology was consistent with this vision, al-though he wrote little on the topic.[19]

As the examples of Levinthal and Finkelstein illustrate, some Conservative rabbis talked like contemporary Reform rabbis when discussing evolution. At the same time, a different approach emerged in the writings of Mordecai Kaplan. A faculty member at JTS, Kaplan was convinced that existing Jewish movements were inadequate to ensure the survival of the Jewish people and Judaism in the face of the challenges of modernity. In response, he called for the "reconstruction" of Judaism and developed a vision of Judaism as an evolving religious civilization: a vision that emphasized belonging over believing, that talked about folkways rather than divine commandments, and that discussed God in functional terms in relation to human needs.[20] No doubt, Kaplan's understanding of evolution as progressive change infused his reading of Jewish history and his judgments about the future of Judaism, but this was not unique to him.[21] What differentiated Kaplan from many Conservative and Reform rabbis was his naturalistic

18. Levinthal, *Judaism*, 33–43, 69–87.

19. Arthur H. Compton, *The Religion of a Scientist* (New York: Jewish Theological Seminary, 1946; originally published 1938); Louis Finkelstein, "Tradition in the Making" [1934], in *Tradition and Change: the Development of Conservative Judaism*, ed. Mordecai Waxman (New York: Burning Bush Press, 1958), 187–97, on 189; Fred Beuttler, "For the World at Large: Intergroup Activities at the Jewish Theological Seminary," in *Tradition Renewed*, ed. Jack Wertheimer, 2 vols. (New York: Jewish Theological Seminary, 1997), 2:669–735, on 675.

20. Mordecai M. Kaplan, "What Is Judaism?," *Menorah Journal* 1 (1915): 309–18; Kaplan, *Judaism as a Civilization* (New York: Reconstructionist Press, 1957; originally published 1934); Mel Scult, *Judaism Faces the Twentieth Century: A Biography of Mordecai M. Kaplan* (Detroit: Wayne State University Press, 1993).

21. Jacob J. Staub, "Evolving Definitions of Evolution," *Reconstructionist* 61, no. 2 (1996): 4–13.

explanation for the origin of religion and his disdain for rational arguments to prove God's existence and attributes. These views in turn shaped his discussions of evolution.

In particular, Kaplan discussed the mental and social evolution of animals and humans in order to explain the origin of belief in God and religion and to substantiate his view that human beings are unique. Kaplan believed that social animals had what he called a "sub-religion" based on "socially beneficent" emotions that "transcend the self." At the same time, he argued that science had discovered important differences between animals and humans: animal evolution is directed by natural selection and leads to adaptation to particular environments, whereas human evolution is directed by intelligent choice and is characterized by adaptability. Most importantly, humans alone possess personality, a hierarchy of meanings and purposes that result from an integration of memory, imagination, and reason, mediated by speech. Only with the development of personality did the idea of God emerge, and only with the development of complex social organization did religion appear. This naturalistic approach to the origin of belief in God and religion differentiated Kaplan from Cohon. For Cohon, human uniqueness ultimately resided in the human soul, which had its source in a transcendent God. For Kaplan, human uniqueness resided in personality and social organization, both of which could be explained by scientific theories of human evolution.[22]

The most striking difference between Kaplan and Cohon, indeed between Kaplan and all other rabbis I have discussed, is his disregard for natural theological arguments that lead from evolution to God. This disregard reflected Kaplan's disdain for traditional theology and metaphysics. While he certainly believed that evolution was a progressive and somewhat open-ended process that included roles for both natural selection and divine processes,[23] Kaplan did not argue from evolutionary biology or physics to the existence or attributes of God. His arguments started with human experience, not a more general understanding of the natural world. As he explained in *Judaism as a Civilization* (1934), the reality of God is "present in the very will-to-live, the reality of which we experience in every fiber of our being." The question that concerned Kaplan was how to draw inferences from individual experience to

22. Mordecai M. Kaplan, "The Relation of Religion to Social Life," *The S. A. J. Review* 6, pt. 36 (1927): 5–12; Kaplan, *Judaism as a Civilization*, 109; Kaplan, *The Meaning of God in Modern Jewish Religion* (New York: Jewish Reconstructionist Foundation, 1947; originally published 1937), 85–86, 111–12, 128, 161–62, 189, 309; Scult, *Judaism Faces the Twentieth Century*, 86.

23. Shai Cherry, "Three Twentieth-Century Jewish Responses to Evolutionary Theory," *Aleph: Historical Studies in Science and Judaism* 3 (2003): 247–90, on 263–78.

cosmic reality. Here he turned not to evolution but to the concept of organism, which had permeated many disciplines in the 1920s and '30s and was central to Alfred North Whitehead's philosophy. As Kaplan wrote, the universe is an "organic totality," and so what is known about a part of the universe—human beings—contains information about the nature of the whole.[24]

Many of Kaplan's followers shared his disregard for traditional theology, and as a result references to evolution were virtually absent from their writings. There were exceptions, such as Eugene Kohn, who argued against a materialist worldview by invoking evolution to support purpose and divine presence within the natural world.[25] However, among Kaplan's followers a radical break soon occurred with regard to their views about God and theology. Milton Steinberg led this revolt in the early 1940s arguing, contra Kaplan, that religion must provide an "interpretation of reality as a whole," one of the traditional tasks of theology. Moreover, his new theology involved a role for evolution. Steinberg adopted an approach similar to that taken by Compton in his 1938 lecture at JTS, although it is not known if any direct influence existed. Both argued that the "hypothesis" of God provided a "more reasonable" explanation of the world than any other hypothesis; hence, belief in God was a reasonable faith, at least as reasonable as belief in any scientific theory. Moreover, like Compton, Steinberg pointed to the "creative" evolution of "novelties" and the presence of "purposive" behavior of organisms as aspects of the natural world, confirmed by science, that the God hypothesis could best explain.[26]

World War II, the Holocaust, and the Problem of Evil

The confrontation of Jewish theologians and rabbis with World War II, the Holocaust, and the problem of evil would eventually help to transform Jewish thinking about God and human nature. However, in the 1940s, many rabbis grappled with the meaning of these events from within their existing religious frameworks. This was the case with Steinberg, Cohon, and Kaplan, who, despite their differences, sought to integrate elements from the scientific

24. Kaplan, *Judaism as a Civilization*, 309–10, 315–16; Kaplan, *Communings of the Spirit: The Journals of Mordecai M. Kaplan, Volume 1: 1913–1934*, ed. Mel Scult (Detroit: Wayne State University Press, 2001), 376–77, 395.

25. Eugene Kohn, *The Future of Judaism in America* (New Rochelle, NY: Liberal Press, 1934), 134–39.

26. Milton Steinberg, "Toward the Rehabilitation of the Word 'Faith' " [1942], in *Anatomy of Faith*, ed. Arthur A. Cohen (New York: Harcourt, Brace, 1960), 63–79; Steinberg, "The Common Sense of Religious Faith" [1947], in ibid., 80–108.

study of evolution into Jewish thought. While Cohon and Kaplan had previously discussed the aggressive and competitive side of human nature with reference to human evolution, only in the 1940s did they, along with Steinberg, focus their attention on the problem of evil.

With World War II raging and the destruction of European Jewry underway, Steinberg turned to grapple with the existence of evil in a world created by God, but without denying either the reality or the tragedy of evil. Viewing evolution as progressive change, in which the higher species retains attributes that had proven useful in previous generations, Steinberg wrote that evil is "the persistence of the circumstances of lower strata in higher." For example, "sullen ferocity" may be appropriate to the gorilla, but is considered "sin" when it appears among human beings. Equally important, evolution is "the saga of life's continuous victory" over survivals from our prehuman past. Human beings, with their intellect and moral insight, can then continue that saga and "subdue what remains of the inner beast." Evolution is directed towards greater "reason, freedom, creativity, and compassion," values that were "written into the scheme of things." Steinberg concluded with the theological reflection that if evolution is the "progressive incarnation of a cosmic Thought-Will," then evil is "the still unremoved scaffolding of the edifice of God's creativity" as God evolves higher forms of being. Evil remains, but new powers emerge in more highly evolved beings that can subdue and direct the evil impulses toward the good.[27]

After World War II, Cohon formulated his own evolutionary theodicy. In *Judaism as a Way of Life* (1948), he argued that the "essential optimism" of medieval Jewish philosophy, which Cohon accepted, "require[d] restatement in the light of evolutionary doctrine." In an evolutionary world, where things are developing toward perfection, evil is "the inevitable concomitant of cosmic experimentation." Natural selection proceeds apace, and that which survives is more "fit to serve the cosmic ends." Viewed from within the finite, there is evil; from the eternal point of view, evil is "the way to and condition of the good."[28]

Kaplan vehemently rejected this approach to evil, but not because he was more pessimistic than Steinberg or Cohon. Rather, Kaplan maintained that our reasoning powers are too weak to enable us to understand why God allows evil to exist, and he feared that any justification of evil would lead to

27. Milton Steinberg, "God and the World's Evil" [1943], in *A Believing Jew: The Selected Writings of Milton Steinberg* (New York: Harcourt, Brace, 1951), 13–31, on 26–31.
28. Samuel S. Cohon, *Judaism as a Way of Life* (Cincinnati: UAHC, 1948), 50–53, 63. For the history of evolutionary theodicy among Reform rabbis see Kohler, *Jewish Theology*, 149–51.

inaction rather than to a fight against evil. In *The Future of the American Jew* (1948), Kaplan discussed the evil of World War II and the Holocaust, taking a sociological rather than a theological perspective. In this context, evolutionary theory had a different role. The problem was not the "destructive impulses" that humans have inherited from their animal ancestors, but the ideologies that remove the constraints of reason and morality and let loose bestial impulses. According to Kaplan, Friedrich Nietzsche was responsible for transforming Darwin's theory into the social "doctrine of natural selection" that ultimately led to the emergence of Nazi ideology and the destruction that followed. Judaism, by contrast, teaches the "doctrine of spiritual selection," which "sees in man's reason and ethical aspirations the evidence of a different destiny from that of the other species." With the emergence of human beings, there is a "reversal of the direction in which nature seem[ed] to be operating" and this new direction—toward reason, equality, and peace—is possible as along as humans choose to follow the teachings of Judaism and act on the "truth that man's survival depends on his transcending the law of natural selection, and subjecting himself to the law of spiritual selection."[29] As his reference to Darwin and Nietzsche suggests, the problem lay not with natural selection as a biological theory, but with its deployment as the foundation for a social ideology.

The Postwar Revolution within American Jewish Thought

The 1950s and '60s saw a revolution in American Jewish thought led by Emil Fackenheim (Reform), Abraham Joshua Heschel (Conservative), and Will Herberg (Marxist turned neo-Orthodox theologian and social thinker). These thinkers argued that World War II and the Holocaust undermined the "optimistic liberal beliefs concerning progress, reason, and human perfectibility" that they believed had shaped Reform, Conservative, and Reconstructionist understandings of Judaism.[30] Each drew on developments in twentieth-century thought, including Protestant neo-orthodoxy, existentialism, and the philosophies of Martin Buber and Franz Rosenzweig. Heschel also drew upon Hasidism and Jewish mysticism.

These theologians believed that an excessive reliance on reason and science had distorted the proper understanding of Judaism and distanced individuals from a personal relationship with the transcendent, living God. Although they tended to speak about science in general terms, Herberg made

29. Mordecai M. Kaplan, *Future of the American Jew* (New York: Macmillan, 1948), 246–56, 317.
30. Sarna, *American Judaism*, 280; see also Goldy, *Emergence of Jewish Theology.*

several references to evolution in his *Judaism and Modern Man* (1951) and identified two sets of problems endemic to theologies that relied heavily on evolutionary ideas. The first set consisted of logical problems. The inference from "the course of evolution" to God faltered because Hume, Kant, and the logical positivists had demonstrated the limits of reason; thus, any claims about God's role in nature were really not derived from scientific observations. In addition, to infer moral principles from the course of evolution violated the naturalistic fallacy. The second set of problems related to authenticity. Was the God derived from evolution a "metaphysical principle" or the living God of Judaism? Were conceptions of God as a process or as a finite being compatible with Jewish views about prayer and revelation? Herberg, as well as Heschel and Fackenheim, argued that any theology that looked to science for concepts about God, human nature, or morality was inauthentic and a distortion of Judaism.[31] For these theologians, science asks different questions than theology and possesses different goals, methodologies, and conceptual frameworks. Science may raise questions that only faith can answer, but it was critical to clearly demarcate the boundaries between science and religion. When these theologians discussed creation, they focused on metaphysical issues regarding the relationship of God to the world, and not on the scientific study of evolution. For them theology should not aim to coordinate faith and reason. Instead, through study of Jewish texts and traditions, it should seek to understand the meaning of the divine-human encounter in Judaism.[32]

While Kaplan, too, had rejected some of the ways science had been used to understand God, the critique of the new theologians was much deeper and had a wider impact. Because science no longer had a constructive role to play in Jewish theology, evolution was simply ignored among the growing number of rabbis who adopted this new approach. An exception was W. Gunther Plaut, who explored in detail the relationship of Judaism to several sciences, including evolutionary biology, in *Judaism and the Scientific Spirit* (1962).

31. Will Herberg, *Judaism and Modern Man: An Interpretation of the Jewish Religion* (New York: Atheneum, 1977; originally published 1951), 34–35, 89–90, 116, 199–203. See also Emil L. Fackenheim, "Can We Believe in Judaism Religiously?," *Commentary* 6 (1948): 521–27; Abraham Joshua Heschel, *Man Is Not Alone* (New York: Harper and Row 1951), 51–56, 105–6.

32. Emil L. Fackenheim, "Apologia for a Confirmation Text," *Commentary* 31 (1961): 401–10; Fackenheim, *Paths to Jewish Belief* (New York: Behrman House, 1960), 19–31, 43–50; Heschel, *Man Is Not Alone*, 30; Heschel, *God in Search of Man* (New York: Farrar, Straus, Giroux, 1955), 15–20; Arnold Jacob Wolf, *Challenge to Confirmands: An Introduction to Jewish Thinking* (New York: Scribe, 1963), 114–18.

Rabbi at Mount Zion Temple in St. Paul, Minnesota, Plaut was a strong advocate of the new theology. He participated in a group of rabbis who met during the late 1950s to discuss theological trends, and he was a proponent of the new "covenant theology" that many Reform rabbis adopted in the 1960s.[33] In *Judaism and the Scientific Spirit*, Plaut argued that religion and science were separate and conflicted only when either overstepped "its legitimate bounds." In addition, the appropriate response to any conflict was to identify the boundary violations and reestablish the proper relationship between faith and science by asserting the legitimate claims of each.[34]

In his chapter on evolution, Plaut maintained that science can legitimately claim that "life evolved from simpler to more complex forms and that increased consciousness (or awareness) seems to be an aspect of increased complexity." But science possesses no authority to decide the "why and wherefore" of evolution. Plaut described three possible options—materialism (natural selection acting without purpose or goal), vitalism (e.g., Henri Bergson's élan vital), and finalism (evolution the result of a "grand design" toward "purposed goals"). For Plaut, this decision was a matter for faith, not science, and he concluded that Jews are finalists because "we believe life to be informed by purposes and goals, which are of God."[35] The separation of science and religion, however, did not mean that science and religion could not point to the same truth. Plaut noted that some scientists followed a theory popularized by the Russian anarchist and social thinker Peter Kropotkin, that cooperation outweighs individual competition as a factor in the evolution of life.[36] He concluded that evolution possesses an "ethic [that] runs parallel to all religious teaching." Moreover, "all evolutionary theorists agree with religion" that humans have a unique attribute, which science labels mind or culture and religion calls spirit. However, these similarities did not mean that Judaism should turn to science for support and insight into its faith claims. Rather, agreement between science and religion suggested that God intended to teach humans the same truth through both science and religion.[37]

33. W. Gunther Plaut, *Unfinished Business: An Autobiography* (Toronto: Lester and Orpen Dennys, 1981), 181–82, 213; Meyer, *Response to Modernity*, 360–64.

34. W. Gunther Plaut, *Judaism and the Scientific Spirit* (New York: UAHC, 1962), 10–13, 20.

35. Plaut, *Judaism and the Scientific Spirit*, 24–26, 46–53. Plaut adopted these options from American paleontologist George Gaylord Simpson's *The Meaning of Evolution* (New Haven: Yale University Press, 1949), 123–29.

36. Gregg Mitman, *The State of Nature: Ecology, Community, and American Social Thought, 1900–1950* (Chicago: University of Chicago Press, 1992), 66–67, 146–201.

37. Plaut, *Judaism and the Scientific Spirit*, 53–56.

Defending Rational Jewish Theology and Challenging the Evolutionary Synthesis

The new movement in Jewish theology did not go unchallenged. A number of rabbis worked to preserve what they considered a more rational approach to theology in which science had a constructive role to play. Moreover, they argued that the new theology represented "obscurantism" and a "failure of nerve" that would lead to a "retreat from the urgent challenges of the here and now." Therefore, they continued to look to the scientific study of evolution as an important component of a reasonable faith.[38] Because of the controversy that ensued, as well as the general "revival of religion" in postwar America, the number of articles and books published on Jewish theology climbed during the 1950s and '60s.[39] Among those who defended the centrality of reason and science in Jewish theology, three stand out because of their interest in and engagement with the scientific study of evolution: Reform rabbis Levi Olan and Roland Gittelsohn and Conservative rabbi Robert Gordis.[40]

These rabbis had to contend not only with new developments in Jewish theology, but also with the development of the evolutionary synthesis within the field of evolutionary biology. Beginning in the 1930s and culminating in the 1950s, scientists from several disciplines developed an evolutionary theory based on population genetics that they believed could explain the origin of species and the broad patterns of evolution. The evolutionary synthesis was thoroughly naturalistic, rejecting as unscientific any explanations that appealed to vital forces or divine powers. Natural selection became the central and primary process in evolution, although other processes such as genetic drift had important roles to play.[41] Because of this transformation, the evolution of life could be more easily

38. Eisendrath, "Can Faith Survive?," 301.

39. Goldy, *Emergence of Jewish Theology*, 29–42; Sarna, *American Judaism*, 274–82.

40. In "What is Our Human Destiny?," *Judaism* 2 (1953): 195–203, and "The Metamorphosis of Man?," *Judaism* 2 (1953): 307–15, Kaplan for the first time presented arguments from the scientific study of evolution to the presence of a "trans-human cosmic force producing moral development." However, other Reconstructionist rabbis followed Kaplan's earlier rejection of natural theology and pursued philosophical analyses of human nature in their arguments for belief in God. See Eugene Kohn, *Religion and Humanity* (New York: Reconstructionist Press, 1953), 13–26; Jack J. Cohen, *The Case for Religious Naturalism: A Philosophy for the Modern Jew* (New York: Reconstructionist Press, 1958), 127–62.

41. Stephen Jay Gould, "The Hardening of the Modern Synthesis," in *Dimensions of Darwinism: Themes and Counterthemes in Twentieth-Century Evolutionary Theory*, ed. Marjorie Grene (Cambridge: Cambridge University Press, 1983), 71–93; William B. Provine, "Progress in Evolution and Meaning in Life," in *Evolutionary Progress*, ed. Matthew Nitecki (Chicago: University of Chicago Press, 1988), 49–74.

interpreted as a mechanistic, materialistic process. Thus, in order to preserve a role for God in evolution, rabbis now turned a more critical eye on natural selection than they had done previously.

At the same time, and to the benefit of those rabbis struggling to relate evolution to Judaism, there was no consensus among the scientists involved in the evolutionary synthesis over the ultimate meaning of evolution. Of particular concern were the views of George Gaylord Simpson, a leading American paleontologist and one of the central architects of the evolutionary synthesis. In *The Meaning of Evolution* (1949), Simpson offered a materialistic understanding of evolution: "[E]volution is a process entirely materialistic in its origin and operation . . . purpose and plan are not characteristic of organic evolution and are not a key to any of its operations."[42] While such provocative statements attracted the attention of many rabbis, Simpson's interpretation of the meaning of evolution was challenged by philosophers and scientists, including his close colleagues, the population geneticist Theodosius Dobzhansky and biologist Julian Huxley.[43] Because of these disagreements the rabbis could turn to any of several scientific authorities for interpretations of evolution that were more congenial to faith in a divine power operating in nature, and they could dismiss those writers whose views were less congenial.

This was especially important for Levi Olan, who after World War II became a leading advocate for a Jewish theology that drew upon science and philosophy. Rabbi at Temple Emanu-El in Dallas, he often presented papers at CCAR annual meetings and became its president in 1967. He believed that the two World Wars, Freudian psychology, and neo-Orthodox theology dealt a blow to the core belief among religious liberals that moral progress would necessarily result from the existence of both reason and freedom. The challenge was to formulate a theology that recognized both good and evil in human nature and provided a warrant for optimism. While he would eventually select Whitehead's philosophy as his preferred metaphysics, in the 1950s and '60s Olan was more eclectic and looked to Bergson, Whitehead, Wilfred Sellars, Conwy Lloyd Morgan, and other contemporary philosophers and scientists for inspiration.[44] Olan argued that a finite, evolving God, rather than an omnipotent God, was more in line with the presence of evil in an evolving

42. Simpson, *Meaning of Evolution*, 291–92; Simpson, *This View of Life: The World of an Evolutionist* (New York: Harcourt, Brace and World, 1964).

43. Michael Ruse, *From Monad to Man: The Concept of Progress in Evolutionary Biology* (Cambridge: Harvard University Press, 1997), 328–38, 385–401, 423–49.

44. Meyer, *Response to Modernity*, 361; Levi A. Olan, "On the Nature of Man," *CCAR Yearbook* 58 (1948): 255–71.

cosmos. Moreover, science had revealed transcendence and creativity as fundamental features in the evolution of life. On this later point, Olan turned for support to Dobzhansky, who had argued that evolution by natural selection resulted in increasing freedom and moments of transcendence (such as the origin of life and the origin of human beings). For humans this involved the ability to transcend—through conscious choice and culture—the bonds of heredity that constrain other animals.[45] Olan argued, like Steinberg, that God was a "more rational explanation" for this view of evolution than an explanation based on purposeless natural selection.[46] Moreover, an evolutionary process marked by creativity and transcendence provided grounds for optimism, even though Olan maintained that such optimism ultimately rested not on science but on a faith that God was "the creative urge in life, which moves toward some meaningful and hopeful end."[47]

Of all Reform rabbis, Roland Gittelsohn was the most interested and concerned about recent developments in evolutionary biology. Rabbi at Temple Israel in Boston, he held leadership positions in the Reform movement and advocated tirelessly what he called "religious naturalism." Gittelsohn believed faith involved more than what science could prove and he aimed to provide "a new vocabulary for ancient and eternal truths." At the same time, he maintained that Jews "could not believe in God unless reason and experience convinced us of His existence" and that evolution provided "the most convincing evidence" for belief in God.[48] To argue for this, Gittelsohn challenged the explanatory power of natural selection and the materialist interpretation of evolution popularized by Simpson. In *Man's Best Hope* (1961) he criticized natural selection by offering several classic arguments regarding the supposed insufficiency of adaptation: for example, new species are initially less well adapted than their immediate ancestors; and some attributes seem to evolve before they are needed.[49] Gittelsohn then turned to the question of

45. Levi A. Olan, "New Resources for a Liberal Faith," *CCAR Yearbook* 72 (1963): 226–39; Theodosius Dobzhansky, *The Biological Basis of Freedom* (New York: Columbia University Press, 1955), 108, 130–35.

46. Levi A. Olan, "Are Religious People Fooling Themselves?" [1966], in *Maturity in an Immature World* (New York: Ktav, 1984), 252–57, on 256.

47. Levi A. Olan, "Is Anything Too Hard for God?" [1951], in *Maturity in an Immature World* 31–36, on 35. Conservative rabbi Jacob Kohn also turned to Bergson and Whitehead to help interpret evolution in theological terms: Jacob Kohn, *Evolution as Revelation* (New York: Philosophical Library, 1963).

48. Roland B. Gittelsohn, *Little Lower than Angels* (New York: UAHC, 1955), 124; Gittelsohn, *Man's Best Hope* (New York: Random House, 1961), 9–10.

49. Gittelsohn, *Man's Best Hope*, 27–36.

God. Simpson had argued that while overall evolutionary trends existed, a detailed analysis of evolutionary patterns indicated that natural selection was sufficient to explain the data. Gittelsohn disagreed. Relying on Julian Huxley, Teilhard de Chardin, and others, he identified five trends in the evolution of life—increases in organization, cooperation, individualization, freedom, and spirit—and concluded that they could best be explained by postulating an immanent, creative, and purposive power. For Gittelsohn, that power was God.[50] Following Huxley he then argued that evolutionary trends could serve as the foundation for ethics. God had moved the evolution of life in certain directions prior to the appearance of human beings; therefore, if humans wanted to be God's partners and promote God's purposes, then the best source for understanding those purposes and the best guide for human behavior was the scientific understanding of evolutionary trends.[51]

Among Conservative rabbis, Robert Gordis, professor of biblical studies at JTS and a leader in the Conservative movement, devoted the most attention to evolution. Gordis shared Steinberg's view that religion should provide a comprehensive philosophy of life that would include an important role for "the conclusions and methods of science," although this role was limited because scientific knowledge was specialized and could not legitimately address issues of meaning and value. In addition, scientists often disagreed about the ultimate meaning of their theories. Yet, as Gordis explained in *A Faith for Moderns* (1960), the scientific study of evolution had discovered a set of facts about the history of life that could best be explained by appeal to God's power. Like Gittelsohn, Gordis was concerned about the growing explanatory power of natural selection and the materialist view of evolution that had been popularized by Simpson. To weaken the explanatory power of natural selection Gordis gleaned from scientific sources several features of life's evolution that he believed could not be explained by natural selection: for example, evolution involves an immense amount of change in a relatively short geological time; the emergence of "something new and unexpected" at each evolutionary stage; the "simultaneous change" of many parts as humans became bipedal; and, according to fossil evidence, "a steady progress in one direction" or "orthogenesis" in many lineages. According to Gordis the scientific evidence shows that evolution is a "ladder from amoeba

50. Ibid., 31–54; Julian Huxley, *Evolution in Action* (New York: New American Library, 1953), 98–117; Pierre Teilhard de Chardin, *The Phenomenon of Man* (New York: Harper, 1959). See also Gittelsohn, *Wings of the Morning* (New York: UAHC, 1969), 68–85.

51. Gittelsohn, *Man's Best Hope*, 88–104. See also Gittelsohn, "God's Voice . . . Or My Own?," *Journal of Reform Judaism* 30 (1983): 65–71.

to man," characterized by ever-greater complexity, efficiency, and consciousness and reaching "the maximum of self-awareness in man." He then posed Steinberg's question in order to challenge Simpson's position: was natural selection or divine power the best explanation of the facts? Gordis answered that from a "tentative reading," the conclusion that a "great Power" created and directed the course of evolution was "far more compatible with the evidence."[52]

While Gordis, Gittelsohn, and Olan published occasionally on evolution during the next few decades, few rabbis appear to have joined their ranks.[53] By the 1980s the message that Jews need not turn to science to understand and support Jewish belief had spread among Reform, Conservative, and Reconstructionist rabbis. This shift may have resulted in part from shortcomings in the defense of a scientifically grounded theology. When asked at the 1964 CCAR convention whether his views had been shaped by prior faith commitments, Gittelsohn responded that even if he had "begun with no faith," science and observations of nature would have led him to his current views; but many found this hard to believe.[54] And for those interested in scientific details, the arguments by Gittelsohn and Gordis against natural selection would have been unconvincing, since they occasionally misrepresented scientists' views and relied upon dated arguments.[55] While these weaknesses may have been important for some, the growing conviction that evolution (and science more generally) was irrelevant to Judaism, and so could be ignored, was certainly due primarily to broader social and intellectual developments: greater acceptance of the new theology; more focus on Israel and

52. Robert Gordis, *A Faith for Moderns* (New York: Bloch, 1960), 15–24, 71–75, 86, 94–98. Conservative rabbi Ben Zion Bokser presented views similar to Gordis's in *Judaism: Profile of a Faith* (New York: Knopf, 1963), 23–40, 61–62, 119–22.

53. Robert Gordis, "The Impact of Science: Religion and Science," *Midstream* 29 (August–September 1983): 26–32; Roland B. Gittelsohn, "Why I Am a Religious Naturalist," *Reform Judaism* 20, pt. 2 (winter 1991): 22–23; Levi A. Olan, *Prophetic Faith and the Secular Age* (New York: Ktav, 1982). See references to evolution in Harold S. Kushner, *When Bad Things Happen to Good People* (New York: Schocken Books, 1982), 51–55, 66.

54. Roland B. Gittelsohn, "No Retreat from Reason!," *CCAR Journal* 74 (1965): 191–215, on 213. For criticism of the use of science by Kaplan, Gittelsohn, and Olan, see Eugene Borowitz, "Faith and Method in Modern Jewish Theology," *CCAR Yearbook* 73 (1963): 215–28.

55. Two examples illustrate the problem. Gittelsohn, *Man's Best Hope* (1961), 34–35, challenged natural selection, citing a 1944 text revised by the author in 1959; compare William Howells, *Mankind So Far* (Garden City, NY: Doubleday, Doran, 1944), 6, with Howells, *Mankind in the Making* (Garden City, NY: Doubleday and Company, 1959), 19. In 1960, Gordis cited a book on the evolution of the kidney as evidence of design, but the author had explicitly argued against this; compare Homer W. Smith, *From Fish to Philosopher* (Boston: Little, Brown, 1953), 210–11, with Gordis, *Faith for Moderns*, 97.

the Holocaust in Jewish thought; emphasis on particularistic concerns in the Jewish community; and growing disillusionment with the objectivity and authority of science in American culture.[56] In turn, developments in American society and Jewish theology would eventually provide an impetus for renewed interest in the topic of evolution.

Scientific Creationism, the Christian Right, and Reform Rabbis

A burst of interest in evolution occurred in the early to mid-1980s, when the scientific creationists' campaign to gain equal time in public schools for the teaching of biblical Creation alongside the scientific theory of evolution was gaining ground. As in the 1920s, only the Reform movement solicited papers and discussed the topic at its annual rabbinic conventions and, as before, Reform rabbis were deeply concerned about the broader cultural agenda of the creationists. The 1980 report of the CCAR Committee on Church and State explained: "The radical right has emerged as a potent force in American politics. . . . Such issues as abortion rights, school prayer, [and] the teaching of 'scientific creationism' . . . symbolize a real and present danger to the pluralistic basis of the America we cherish."[57] At the 1982 CCAR convention Rabbi William Leffler discussed his efforts to organize resistance to scientific creationism in the public schools of Lexington, Kentucky. The CCAR executive board also adopted a resolution challenging the creationists' claim for "balanced time" on the grounds that it violated the principle of the separation of church and state. In 1984 a resolution on the same topic was adopted by the CCAR as a whole. The CCAR also directed the movement's Religious Action Center in Washington, D.C., to build coalitions to fight the rise of the Christian right.[58]

While Reform rabbis had a strong interest in the legal and political issues, they expressed comparatively little interest in the relationship between evolution and Judaism. In contrast to what transpired in the 1920s, there is no record of discussion on this topic at the CCAR conventions. However, in 1982 the *Journal of Reform Judaism* published two articles on the topic of Judaism,

56. Sarna, *American Judaism*, 306–18, 333–38; Edward A. Purcell, Jr., "Social Thought," *American Quarterly* 35 (spring/summer 1983): 80–100.

57. Frank N. Sundheim, "Report of the Committee on Church and State," *CCAR Yearbook* 90 (1981): 34–33.

58. Frank N. Sundheim, "Report of the Committee on Church and State," *CCAR Yearbook* 91 (1982): 28–29; Sundheim, "Report of the Committee on Church and State," *CCAR Yearbook* 92 (1983): 127–28; Sundheim, "Scientific Creationism," *CCAR Yearbook* 92 (1983): 313–14; William J. Leffler, "The Problem of Scientific Creationism," *CCAR Yearbook* 92 (1983): 97–98; "On Creationism in School Textbooks," *CCAR Yearbook* 94 (1985): 151.

evolution, and creationism, one by Leffler and the other by Rabbi Jack Luxemburg. Their positions illustrate just how far many Reform rabbis had traveled since the 1920s. Leffler maintained that "we need to delineate and emphasize the differences between the methodologies of religion and science."[59] Luxemburg emphasized the "scope and limitations" of science and religion, indicating that "it is inappropriate for science to attempt either to prove or disprove the existence of God."[60] These articles illustrate that by the early 1980s most American rabbis had accepted the separation of science and religion as postulated by the new theology.[61] Moreover, while paying little attention to the details of evolutionary theory, Leffler and Luxemburg linked their position to their political agenda: scientific creationism was illegitimate because it violated the boundaries between science and religion. In turn, the independence of religion and science supported the doctrine of separation of church and state.

Jewish Mysticism, Kabbalah, and a New Story of Creation

Despite the apparent divorce between science and religion in the 1980s, evolutionary ideas again reentered Jewish theology as part of a renewed interest in Jewish mysticism, especially kabbalah.[62] The initial focus was on Gaia and cosmic evolution. In *The River of Light: Spirituality, Judaism, and Consciousness* (1991), Lawrence Kushner, a Reform rabbi and author of several popular books on Jewish mysticism, explored how the Gaia hypothesis and modern theories of cosmology might illuminate God's relationship to the evolution of consciousness. Likewise, Rabbi Zalman Schachter-Shalomi, leader of the Jewish Renewal movement, which influenced rabbis from across the spectrum of American Jewry, began to incorporate ideas about Gaia and cosmic evolution into his teachings.[63] But, it was Arthur Green who became

59. William J. Leffler, "Some Insights on Creationism," *Journal of Reform Judaism* 29 (1982): 50–55, on 50.

60. Jack Luxemburg, "Science, Creationism, and Reform Judaism," *Journal of Reform Judaism* 29 (1982): 42–49, on 44, 45.

61. Arnold Eisen, "American Jewry in the Twenty-First Century: Strategies of Faith," in *The Americanization of the Jews*, ed. Robert M. Seltzer and Norman J. Cohen (New York: NYU Press, 1995), 451–57.

62. See Robinson (chapter 3, this volume) for how twentieth-century Orthodox Jews used kabbalah differently, to argue that Judaism transcends the scientific study of evolution.

63. Lawrence Kushner, *The River of Light* (Woodstock, VT: Jewish Lights, 1981), pp. ix–xi, 74–91, 94–110; Zalman Schachter-Shalomi, *Paradigm Shifts* (Northvale, NJ: Aronson, 1993), 149, 269–75, 301–2. See also Daniel C. Matt, *God and the Big Bang* (Woodstock, VT: Jewish Lights, 1996); Lawrence Troster, "From Big Bang to Omega Point: Jewish Responses to Recent Theories in Cosmology," *Conservative Judaism* 49, pt. 4 (1997): 17–31.

particularly interested in infusing ideas derived from evolutionary biology into theology, first in *Seek My Face, Speak My Name* (1992) and most recently in *Ehyeh* (2003).

While Green incorporated aspects of both Heschel's and Kaplan's theology into his own, his general view about the relationship between science and religion was closer to Heschel's.[64] He claimed that science and religion were separate and complementary; science took a "more mechanistic" view of reality, while religion emphasized "consciousness and will."[65] Moreover, Green did not turn to science as a foundation for Jewish faith but instead systematically grounded his faith claims in traditional Jewish sources. At the same time, Green had a deep interest in evolution shaped both by his theology and by his views about the social and moral impact of modern evolutionary theory.

For Green, stories of creation, of origins, are essential in our quest for meaning. However, while adopting many of the insights, principles, values, and perspectives of Jewish mysticism, he could not accept the metaphysical system of sefirot that kabbalists had used to explain "the stages by which God is revealed." Instead, the "bio-history of the universe," became "the only sacred drama that really matters"—the "ongoing account of how Y-H-W-H, source of life, reached forth into the world of form, became manifest in the infinite variety of species, and finally became articulate in the consciousness and language of humanity."[66] Like his predecessors, Green offered a progressive interpretation of evolution as the emergence of ever more complex forms of life culminating, to date, in human beings, whose self-consciousness makes them unique. At the same time, and unlike his predecessors, he emphasized "the divergence of species from one another," connecting this with the kabbalistic notion of the "great striving of the One to be manifest in the garb of the many." Emphasis on divergence formed part of Green's effort to construct a theology of creation that embodied attitudes and values necessary for the preservation of life's diversity.[67]

At the same time, like Kaplan, Green was deeply concerned about how natural selection theory had been used to justify bloodshed and violence in the twentieth century, although he seemed to place a greater responsibility on scientists themselves than had Kaplan. Green explained that a narrative of

64. Arnold Eisen, "Jewish Theology in North America: Notes on Two Decades," *American Jewish Year Book* 91 (1991): 3–33, on 24–27.

65. Arthur Green, *Ehyeh: A Kabbalah for Tomorrow* (Woodstock, VT: Jewish Lights, 2003), 114.

66. Ibid., 23, 111; Green, *Seek My Face*, 54.

67. Green, *Seek My Face*, 54; Green, *Ehyeh*, 113–19.

origins "more or less began with Darwin and is refined daily by the work of life scientists and physicists" in which the "history of living creatures is again depicted as a bloody and violent struggle, the implications of which for human behavior—even for the possibility of ethics—have hardly gone unnoticed." He drew a parallel between this narrative and the prebiblical creation myths that spoke about "conflict, slaughter, and victory" among the gods. As an alternative to the prebiblical myths Genesis had offered a vision of harmony, in which "everything had its place as the willed creation of a single deity and all conflict has mysteriously been forgotten." Similarly, he proclaimed that a new creation narrative was now needed that could provide a vision of harmony, unity, and peace to replace the current emphasis on struggle and conflict.[68]

While Green did not explicitly criticize the evidence and arguments scientists offered for the contemporary theory of natural selection, as had Gittelsohn and Gordis, he did reference a group of contemporary scientists who had done just that and who appeared open to "describing the origin and evolution of species, in some sense, as the expression of a singular universal force . . . that strives relentlessly, though by no means perfectly, toward greater complexity and consciousness." For Green, science was one of several paths toward awareness of God's presence and unity, and he believed that "in a formula not yet articulated" the scientific and religious understandings of reality would meet.[69] However, his reference to scientists who emphasized the evolution of consciousness and downplayed natural selection suggests that Green himself thought that the mainstream scientific theory of evolution by natural selection would have to be modified in order to serve as a source for awareness of God.

At the same time, like many who preceded him, Green did not reject natural selection; in fact, he incorporated natural selection in his discussion of theodicy. Green argued that the "struggle for territory, limited resources, and more fruitful mating partners" explained the "aggressive urges that are indeed a vital part of our inner nature," the "evil urge" in each person. Green then provided an "evolutionary version" of the kabbalists' belief that the

68. Green, *Ehyeh*, 109–11; Green, *Seek My Face*, 54–55, 220–21.

69. Green, *Seek My Face*, 54, 219; Green, *Ehyeh*, 32, 114. Green mentions the physicist Brian Swimme by name in Green, "A Kabbalah for the Environmental Age," in *Judaism and Ecology*, ed. Hava Tirosh-Samuelson (Cambridge: Harvard University Press, 2002), 8. For a collection of essays advancing a new view of evolution, which includes chapters by Swimme and also the biologist Charles Birch, see David Ray Griffin, ed., *The Reenchantment of Science: Postmodern Proposals* (Albany: SUNY Press, 1988).

presence of the demonic in nature and in human beings is a necessary part of the process by which all of created life came into existence. The scientific theory of evolution tells us that God "had to emerge" in a way that made aggression, struggle, and evil constituent parts of reality.[70] Thus Green followed Steinberg and Cohon by turning to evolutionary theory to help reformulate traditional Jewish ideas.

Evolution as a Theological, Social, and Political Resource

Most Reform, Conservative, and Reconstructionist rabbis who viewed science as relevant to Jewish faith offered a variety of arguments showing that evolutionary science strongly implied the need for the everpresent creative power of God: the patterns and products of evolution point to divine intelligence; given the facts about evolution, God is the best hypothesis to explain them; and natural selection is insufficient to explain certain trends or episodes in life's evolution. And while Green avoided such arguments, he hoped that scientific and religious views of evolution would eventually converge. These rabbis also turned to evolution to help explain human nature, focusing primarily on how natural selection explained aggressive and competitive impulses, and for some rabbis this became the basis for an evolutionary theodicy. At the same time, all the rabbis under discussion thought that the scientific study of human beings was of limited scope and that the divine was necessary to understand adequately the most unique and important features of human nature. And since they were able to point to scientists with similar convictions, even if they were not always at the leading edge of research, these rabbis maintained that they were being true to the best of science as well as authentic in Jewish faith.[71]

The need to negotiate a balance between scientific theory and religious belief, as well as the roles played by social and political context in shaping rabbinic responses, is best illustrated by the way rabbis engaged the topic of natural selection. Even before the evolutionary synthesis, many scientists accorded natural selection an important role in explaining the history of life, and the rabbis in this study accepted that: indeed, they seemed to welcome

70. Green, *Ehyeh*, 149–52.

71. Although it would be premature to draw comparisons and contrasts between the rabbis discussed here and their contemporaries among the Protestant or Catholic clergy, Peter Bowler's study of British debates suggests that they had much in common, including criticism of natural selection, preserving a role for God in evolution, and an appeal to scientists to bring credibility to their positions. See Bowler, *Reconciling Science and Religion*, 220, 248–49, 258, 266, 273, 281–85, 299, 395–98.

its explanatory power especially in regard to human nature. But they also recognized that natural selection theory had been used to support both materialist philosophies that threatened Jewish faith (and supposedly congregational affiliation) and social ideologies that justified militaristic and anti-Semitic behavior. Thus, in their rejection of scientism, materialism, militarism, and anti-Semitism, these rabbis limited the role of natural selection in the evolution of life while retaining a role for the divine.[72]

Indeed, when rabbis discussed evolution, they were more concerned with the larger social and intellectual issues at stake than the details of evolutionary theory. Moreover, engaging the scientific study of evolution—at any level of detail—and finding it relevant for Jewish faith was in itself to take a position in debates about the nature of Jewish theology. This position, common prior to the 1050s, was eventually pushed to the sidelines by the growing acceptance of new approaches to Jewish theology and by social and intellectual trends in the larger Jewish and American communities, which tended to set science and religion apart. This strategy became aligned with the separation of church and state, the defense of religious pluralism, and the opposition to teaching scientific creationism in public schools. It also set the stage for Green's call to reintegrate science into Jewish theologies of creation. And now, in the context of renewed controversy in America over the teaching of evolution in public schools and shifts in the Jewish community regarding the principle of separation of church and state, efforts to reconnect evolution with Judaism are again loaded with potential legal and political ramifications.[73] How these will be played out remains to be seen.

72. As suggested in this chapter, I believe that Kaplan was less critical of natural selection than has been argued by Cherry, "Three Twentieth-Century Jewish Responses," 270–77.

73. For example, see the exchange between Rabbi Michael Lerner and historian of American religion Martin Marty in "Interview with Martin Marty," *Tikkun* 20, no. 4 (July–August 2005): 35–39.

3

"Practically, I Am a Fundamentalist": Twentieth-Century Orthodox Jews Contend with Evolution and Its Implications

Ira Robinson

In the summer of 1925, the world's attention was focused on the so-called Monkey Trial in Tennessee, which debated the truth as well as the propriety of teaching the theory of evolution. In Montreal, an immigrant Orthodox rabbi, Hirsh Cohen,[1] wrote to his daughter and son-in-law voicing his opinion:

> [Regarding] the Darrow-Bryan dispute, as long as it is in theory, one can agree with whatever position one thinks right and still remain a believer in the divinity of the Bible. It is the power of the Torah that all theories can be included. When Alexander von Humboldt and other natural scientists discovered that in the earth there are rock formations that were much, much older than our Torah's chronology allows for, the sages of the Torah were not shocked, and they realized that this way of thinking was long known to the sages of the Talmud and the kabbalists . . . that our present world is not the first.[2] . . . However, as I said, this is only in theory. Practically, I am a fundamentalist. Our great rabbi, Maimonides, philosophized in his *Guide of the Perplexed* in many matters theoretically. But when in his *Yad ha-Ḥazakah*[3] he dealt with practical things, he was altogether different.[4]

I would like to acknowledge the helpful comments of Geoffrey Cantor, Marc Swetlitz, and Ronald Numbers on the paper as originally presented at the Arizona State University conference.

1. On Cohen, see Ira Robinson, *Rabbis and Their Community: Studies in the Eastern European Orthodox Rabbinate in Montreal, 1896–1930* (Calgary: University of Calgary Press, forthcoming), ch. 2.
2. For a similar contemporary rabbinic view of the age of the earth, see Abraham Isaac ha-Kohen Kook, *Igrot ha-Re'iyah*, 4 vols. (in Hebrew) (Jerusalem: Mossad ha-Rav Kook, 1985), 1:104; Carl Feit (chapter 9) in this volume.
3. *Yad ha-Ḥazakah* is Maimonides' legal code, otherwise known as *Mishneh Torah*, written in the late twelfth century.
4. The letter is dated 1 Devarim 5685 [19 July 1925], Cohen Papers, Canadian Jewish Congress National Archives, Montreal. On the impact of Darwin on other immigrant Orthodox rabbis, see Kimmy Caplan, *Orthodoxy in the New World: Immigrant Rabbis and Preaching in America, 1881–1924* (in Hebrew) (Jerusalem: Zalman Shazar Center for Jewish History, 2002), 259.

Here Rabbi Cohen underscored the idea that the halakhic (legal) aspects of Judaism were to be taken literally, whereas aggadic (nonlegal) opinions of the rabbis were open to interpretation. As we will see, Cohen's use of this distinction foreshadows some of the problems and strategies adopted by Orthodox Jews in the twentieth century as they contended with the theory of evolution and its implications. Cohen may have encountered "fundamentalist" in popular media presentations of the contemporary public debate over evolution, but he then used the word idiosyncratically in applying the term "practical fundamentalist" to someone who accepts the Torah's legal system as true and valid, whether or not that person also accepts literally the Torah's account of Creation. Many of the Orthodox Jews discussed in this chapter share this "practical fundamentalism" while distancing themselves from Christian fundamentalism.

Although Cohen distinguished the practice of Judaism from attitudes toward evolution, many Orthodox Jews have been troubled by the theory, as a result of the challenge that evolution has mounted to the traditional Jewish myth of Creation. To anthropologists, myths are stories that explain the nature of reality. Yet from this anthropological perspective science also serves as myth. As Colin Grant points out, in modern Western society nothing rivals science "as an official definer of reality."[5] In claiming to explain why things are as they are, science has largely displaced previously dominant myths and, in particular, the myths intrinsic to such scriptural religions as Christianity, Islam, and Judaism.

Throughout much of history, traditional Jews looked to Jewish texts and their interpretation to provide a comprehensive account of reality. Not only have they sought to live their lives in accordance with the divinely ordained precepts to be found in Torah, but over the generations Jews have studied it closely, discussed it in detail and added to the wealth of commentaries. The Torah tradition has often clashed with rival mythic systems. In the present context, the most interesting and significant of these clashes occurred in the medieval period, when Jewish thinkers encountered the legacy of ancient Greek philosophy and science. This rival mythic system seemed to present a comprehensive view of reality that denied the truth of Torah and other divine revelations. Therefore, intellectually aware Jews had to respond to Greek philosophy, particularly in those areas where it was perceived to conflict with the Torah.

Although this chapter cannot do justice to the complexities of the medieval confrontation between Torah and Greek natural philosophy,[6] we

5. Colin Grant, *Myths We Live By* (Ottawa: University of Ottawa Press, 1998), 30, 41.

6. This chapter does not discuss the attempts by Philo of Alexandria to reconcile Torah and Greek thought, since the Orthodox Jews did not have access to Philo's writings, but instead drew extensively on the works of medieval philosophers and rabbis.

can identify four basic strategies adopted by Jewish thinkers. Faced by what they considered to be an irreconcilable conflict, some Jews remained loyal to Torah and rejected outright Aristotelian philosophy as subversive and dangerous. This first stance might be termed the "rejection of science." Thus, while the Torah asserted that God had created the universe, Aristotelian philosophers claimed that the universe was eternal and they thus dispensed with a Creator.[7] Rejectionists simply dismissed such claims and cautioned other Jews to avoid the secular sciences.[8]

Other Jews sought to reconcile the truth of Torah with that of science. This second strategy was founded on the belief that the Torah, when correctly understood and interpreted, makes identical claims to science. This strategy was commonly employed by Saadia ben Joseph Gaon, the leading scholar of rabbinic Judaism of the tenth century. For example, given the claim that God is "one, living, omnipotent, and omniscient, that there is nothing that resembles Him, and that He does not resemble any of His works," Saadia first proceeded to cite biblical verses proving all these assertions. Then, he wrote, "having learned about these five facts from the books of the prophets, we proceeded to confirm them by way of logical reasoning and found them to be correct."[9] We may term this form of reconciliation the "integration of science and Torah."

However, there were occasions when integration could not be accomplished and medieval Jewish thinkers then adopted a third strategy. This was to insist that Greek science and philosophy were fallible. By demonstrating the weaknesses in their opposition's arguments, Jewish philosophers hoped to disparage the Greek tradition and, by implication, strengthen the claims of the Torah. Thus in the twelfth century, Moses Maimonides asserted in his *Guide of the Perplexed* that Aristotelian science had failed to explain correctly the retrograde motions of the planets.[10] This criticism contributed to his attack on the Aristotelian claim that the world is eternal and by showing that this claim was merely probable, Maimonides enabled Jews to retain their

7. Oliver Leaman, "Introduction to the Study of Medieval Jewish Philosophy," in *The Cambridge Companion to Medieval Jewish Philosophy*, ed. Daniel H. Frank and Oliver Leaman (Cambridge: Cambridge University Press, 2003), 3–15.

8. J. Dan, prolegomenon to R. Moshe Taku, *Ketav Tamim: Ms. Paris H711* (in Hebrew) (Jerusalem: Merkaz Dinur, 1984).

9. Saadia b. Joseph, *Book of Beliefs and Opinions*, trans. Samuel Rosenblatt (New Haven: Yale University Press, 1948), 94–95.

10. Moses Maimonides, *Guide of the Perplexed*, trans. Shlomo Pines, 2 vols. (Chicago: University of Chicago Press, 1963), 2:24, 322–27; Tzvi Langermann, "Maimonides and the Sciences," in Frank and Leaman, *Cambridge Companion*, 157–75.

belief in divine Creation.[11] This strategy could be called the "weaknesses and limitations of science."

The fourth strategy was to approach—and transcend—science by means of kabbalah. This strategy of understanding science as related—and subordinate to—kabbalah can be traced in Judaic sources from the sixteenth to the twentieth centuries, and is especially well adumbrated in Rabbi Pinḥas Elijah Hurwitz's late eighteenth-century work, *Sefer ha-Berit*,[12] and in the writings of the Italian rabbi and kabbalist, Vittorio Ḥayim Castiglioni.[13] Thus Hurwitz states that science is useful in that it can confirm the divine wisdom of the ancient rabbis and the wisdom of kabbalah.[14] This strategy can be termed the "transcendence of science."

This chapter will demonstrate that in the past century Orthodox Jews have utilized these traditional strategies in responding to the perceived challenge of the theory of evolution. As Elliot Pines, an electrical engineer who has written on Judaism and science, states: "In essence, none of these [perceived conflicts in regards to evolution] is new. They actually go back to older conflicts arising between Torah and various philosophical systems."[15]

Some Orthodox Responses

In confronting the clash between Torah Judaism and the theory of evolution, many twentieth-century Orthodox Jews are primarily concerned that if evolution were accepted uncritically, Torah would be deemed not only irrelevant but false. Orthodox Jews perceived three critical issues. Their primary concern

11. Though many interpreters assume that Maimonides may have believed in the eternity of the universe, it is important in this instance to assert that most contemporary Orthodox readers of Maimonides understand that his conclusion is that Aristotle did not in fact prove his point beyond the shadow of a doubt. William Etkin claims that "like Maimonides, we may say that where reason fails we must choose to believe in creation by a purposeful Creator for religious reasons"; see Etkin, "Science and Creation," in *Challenge: Torah Views on Science and Its Problems*, ed. Aryeh Carmell and Cyril Domb, 2nd ed. (Jerusalem: Association of Orthodox Jewish Scientists; New York: Feldheim, 1978), 240–53, on 251.

12. Ira Robinson, "Kabbala and Science in *Sefer ha-Berit*: a Modernization Strategy for Orthodox Jews," *Modern Judaism* 9 (1989): 275–88.

13. Lois Dubin, "*Pe'er ha-Adam* of Vittorio Ḥayim Castiglioni: An Italian Chapter in the Jewish Response to Darwin," in *The Interaction of Scientific and Jewish Cultures in Modern Times*, ed. Yakov Rabkin and Ira Robinson (Lewiston, NY: Mellen, 1994), 87–101.

14. Robinson, "Kabbala and Science," 281.

15. Elliot M. Pines, "Torah, Reality, and the Scientific Model" (dated February 2001), www.613.org/torah-pines.html. See also Alvin Radkowsky, "Judaism and the Atomic Age," in *The Jewish Library*, vol. 4, *Judaism in a Changing World*, ed. Leo Jung (London: Soncino, 1971), 237.

has been to refute the view that the theory of evolution makes God irrelevant. As the Israeli microbiologist Morris Goldman stated, "God is irrelevant in the Darwinian evolutionary scheme and that is what is wrong with it for a Jew."[16] Likewise, Rabbi David Gottlieb, who abandoned the pursuit of academic philosophy to teach at the Ohr Somayach Yeshiva, argued that by accepting the theory of evolution Jews are conceding to atheism and "are giving up the whole of life as evidence for God."[17] The second issue was morality, because the theory of evolution is seen as "an egregious blueprint for secular humanism [a] blueprint [that] dismantles social order by tacitly approving a) the abandonment of the Almighty as a moral authority, and b) the intentionally inevitable pursuit of cutthroat behavior."[18] Finally, the theory of evolution challenges the concept of man as a "qualitatively different creation."[19] With these significant issues at stake, Orthodox rabbis, scientists, and laypeople have responded to the theory of evolution by employing exegetical and apologetic strategies very similar to those used by medieval Jewish thinkers who confronted the challenge of Greek philosophy and science.

REJECTION OF SCIENCE

Orthodox Jews can be classified on the basis of their accommodation to secular education and culture.[20] Among those who reluctantly accept the secular education of their children only because of state or national standards and requirements, the "rejection of science" strategy is particularly evident when it comes to those aspects of science that are viewed as incompatible with Torah. Thus one of the most prominent Orthodox rabbinical figures of the past generation, Moshe Feinstein, suggested the following solution when

16. Morris Goldman, "A Critical Review of Evolution," in Carmell and Domb, *Challenge*, 216–34, on 218. Cf. Yocheved Golani, review of Lee Spetner, *Modern Theory of Evolution* (New York: Judaica Press, 1997), http://www.jewishpress.com/page.do/423/Book_Review.html (posted 4 October 2001).

17. David Gottlieb, "The Theory of Evolution" (audiotape) G-98, *Jerusalem Echoes* (Jerusalem: Ohr Somayach International, n.d.). See also Dovid Brown, *Mysteries of the Creation: A Cosmology Derived From Torah, Nevi'im, C'suvim, Mishna, G'morroh, and Midrash* (Southfield, MI: Targum/Feldheim, 1997), 248.

18. Golani, review of Spetner.

19. Anonymous, "The Jewish Prospective on Evolution," www.hanefesh.com/edu/Evolution.htm.

20. Though it claims to be a faithful replication of the premodern rabbinic tradition, Orthodox Judaism is very much a product of the wrenching changes undergone by Judaism in the modern era. See Jacob Katz, *A House Divided: Orthodoxy and Schism in Nineteenth Century Central European Jewry* (Hanover: Brandeis University Press, 1998), 36, 71; Charles Liebman, *Aspects of Religious Behavior of American Jews* (New York: Ktav, 1974), 111–88.

addressing the problem of the presentation of the theory of evolution and the related issue of the age of the universe in secular textbooks:

> Textbooks of secular studies that contain matters of heresy [*kefirah*] with respect to the creation of the world are certainly books of sectarianism [*minut*] that are forbidden to be taught. It is necessary to see to it that the secular studies teachers do not teach from them to students. If it is not possible to obtain other books, it is necessary to tear out those pages from the textbooks.[21]

Indeed, some Jewish schools, particularly ultra-Orthodox ones, follow this advice and tear out textbook pages dealing with the theory of evolution.[22]

Twentieth-century Orthodox attitudes to education are summarized by Gerald Schroeder, an Orthodox Jewish scientist whose writings will be discussed below:

> My son . . . had been taught to relate to the Bible in its most literal sense, and so for him, and for many of his teachers, the age of the universe is exactly the age derived from the generations as they are listed in the Bible. For them, the cosmological estimate of the age of the universe, some 15 billion years, is a preposterous fiction.[23]

This rejectionist strategy also has implications outside the realm of education. For example, there have been protests by ultra-Orthodox Jews against an Israeli company, Tara Dairy, for its use of images of dinosaurs in an advertising campaign.[24] As a North American Orthodox critic of this ultra-

21. Moshe Feinstein, *Igrot Moshe, Yoreh De'ah*, vol. 3, responsum 73 (New York: Noble Press 5742 [1982]), 323. On Feinstein, see Ira Robinson, "Because of Our Many Sins: The Contemporary Jewish World as Reflected in the Responsa of Rabbi Moses Feinstein," *Judaism* 35 (1986): 35–46.
22. Judah Landa, *Torah and Science* (Hoboken, NJ: Ktav, 1991), 290; William Shaffir, "Boundaries and Self-Preservation among the Hasidim: a Study in Identity Maintenance," in *New World Hasidim: Ethnographic Studies of Hasidic Jews in America*, ed. Janet S. Belcove-Shalin (Albany: State University of New York Press, 1995), 31–68, on 46. For Hasidism see Jerome Mintz, *Legends of the Hasidim: An Introduction to Hasidic Culture and Oral Tradition in the New World* (Chicago: University of Chicago Press, 1968), 25–158.
23. Gerald L. Schroeder, *Genesis and the Big Bang: The Discovery of Harmony between Modern Science and the Bible* (New York: Bantam, 1990), 11. See also AOJS Students' Questions Panel, "Actual and Possible Attitudes to Evolution within Orthodox Judaism," in Carmell and Domb, *Challenge*, 279: "This science has been a closed book to the average Orthodox student"; and Selya (chapter 8) in this volume.
24. *Jerusalem Post* (international edition), 21 August 1993. For a similar controversy involving Pepsi Cola's use of the evolution of humans in an advertising campaign see "Creation Science News," http://www.answersingenesis.org/creation/v14/i4/csnews.asp. Also Benjamin Svetitsky, "Cosmology (was: Dinosaurs and Kashrut)" (28 September 1993), http://www.emax.ca/mj_ht_arch/v9/mj_v9i37.html.

Orthodox protest, Rabbi Benjamin Hecht, who directs a Torah institute, has stated:

> Belief in the existence of dinosaurs—with the corollary approval of the theory of evolution—was simply deemed to be . . . sacrilegious. Use of dinosaurs in the campaign implicitly demonstrated acceptance of these irreverent ideas which represented a challenge to the truth of Torah and its declaration of a creation, 5764 years ago, in seven days.[25]

Although those who reject evolution because of its incompatibility with Torah appear to be adopting a position identical to that of many Christian fundamentalists, it is important to note that Orthodox Jews, whether or not they treat the theory of evolution or the related issue of the age of the universe as "a preposterous fiction," are united in their opposition to Christian creationism. Although there are some points that Orthodox Jews and Christian creationists might agree upon—for example, that the Hebrew Bible was revealed by God—these Jews clearly want to distance themselves from the Christian fundamentalists.[26] The main reason for this attitude is that creationism is based on the King James Bible and not on traditional Jewish texts, which incorporate the cumulative perspectives obtained from traditional Torah commentaries.[27]

Fundamentalism and creationism have been decried by Orthodox Jews as "nonsense"[28] and "a grave error."[29] Another Orthodox writer described as "frightening" any suggestion that Orthodox Judaism might be aligned with Christian fundamentalism.[30] Orthodox critics of creationism would agree with American Jewish philosopher Norbert Samuelson's opinion that "even

25. Benjamin Hecht, *Dinosaurs*, Insight 5764–#05 (Toronto: Nishma, 2003). This position was supported by, among others, Rabbi Menachem Mendel Schneersohn, the Lubavitcher rebbe. See Alexander Nussbaum, "Creationism and Geocentrism among Orthodox Jewish Scientists," *Reports of the National Center for Science Education* 22, no. 2 (January–April 2002): 38–43.

26. I have been able to find only one source in which the author specifically identifies herself as a "Creationist," and she too expressed her initial understanding of creationism as the province of "narrow minded, Bible-belt, Christian fundamentalists"; see Sara Yoheved Riegler, "Confessions of a Creationist," www.aish.com/spirituality/philosophy/Confessions_of_a_Creationist.asp.

27. Susan Schneider, "Evolutionary Creationism: Torah Solves the Problem of Missing Links" (1984), www.orot.com/ec.html. Cf. Eliezer Zeiger, "Kosher Evolution," *Points of Wisdom: The Newsletter of the Torah Science Foundation* 3, no. 1 (Tevet 5764/December 2003), www.torahscience.org/newsletter7.html.

28. Gottlieb, "Theory of Evolution."

29. Nosson Slifkin, "Science Wars," www.torah.org/features/secondlook/sciencewars.html. Cf. "Creationism vs. Evolution: Radical Perspectives on the Confrontation of Spirit and Science," *Tikkun*, September–October 1987, 55.

30. Baruch Sterman, "Judaism and Darwinian Evolution," *Tradition* 29 (1994): 48–75, on 70.

when read literally, these revered [biblical] texts do not say what Christian 'creationists' say that they mean."[31] The antagonism of the Orthodox toward fundamentalists may also be attributed in part to the specifically Christian nature of the Creation Research Society, which, after some debate, required its members to acknowledge Jesus Christ as their savior.[32]

WEAKNESSES AND LIMITATIONS OF SCIENCE

Unlike Feinstein, some Orthodox Jews do not wish to ignore or suppress discussion of the theory of evolution within their community, bur rather seek to engage evolution with arguments. Moreover, in having to contest the theory at a time when it is generally accepted by the scientific community, they also have to respond to critics like author A. N. Wilson who claim that they are crackpots.[33] Thus Rabbi Aharon Lichtenstein, a leading thinker in the contemporary Modern Orthodox camp, states that:

> confronted by evident contradiction [between Torah and science], one would . . . initially strive to ascertain whether it is apparent or real . . . whether indeed the methodology of *madda* [science] does inevitably lead to a given conclusion, and . . . whether . . . Torah can be interpreted . . . so as to avert a collision.[34]

Following Lichtenstein's advice to test the conclusions of science, a number of writers have sought to demonstrate defects in the scientific argument for evolution. Some of the arguments against the scientific basis of the theory of evolution are fairly technical in nature, while others are more popular and rhetorical. This strategy is likely to appeal to a wide public, for critics of evolution are riding a wave of skepticism: a recent poll concerning scientific literacy in Quebec indicated that, whereas 94 percent of respondents accept the validity of Einstein's theory of relativity, only half of them accept the theory of evolution as valid.[35]

The scientific arguments are generally made by Orthodox Jews who have received a training in science or engineering. Some of them, but by no means

31. Norbert Samuelson, "The Death and Revival of Jewish Philosophy," *Journal of the American Academy of Religion* 70 (2002): 117–34, on 128.

32. Ronald Numbers, *The Creationists* (New York: Knopf, 1992), 230–31, 318. For evidence of Jewish collaboration with the Creation Research Society, see Moshe Trop, letter to the editor, *Creation Research Society Quarterly* 20 (1983–84): 121–22.

33. A. N. Wilson, *The Victorians* (New York: Norton, 2003), 100.

34. Cited in Sterman, "Judaism and Darwinian Evolution," 49.

35. Allison Lampert, "Biology Teachers Fear Paring Knife," *Gazette* (Montreal), 5 May 2003, A1, A3.

all, have also undertaken rabbinic training. Their growing presence, barely evident early in the twentieth century,[36] increased considerably after the Second World War and culminated in the founding in 1948 of the Association of Orthodox Jewish Scientists (AOJS). One of the association's principal aims was to resolve the "apparent challenges of scientific theory to Orthodox Judaism."[37] In their critiques of scientific theories deemed to challenge Orthodox Judaism, such as the theory of evolution, members of the AOJS engaged basic issues of faith. Thus one of its leading members, the eminent nuclear engineer Alvin Radkowsky, argued that these contradictions involved a test of faith comparable to that of the biblical Abraham, who was called upon to offer his beloved son, Isaac, as a sacrifice. "In these circumstances," he wrote, "the Jewish scientist must hold staunchly to a faith [that] implies that the claims of secularism . . . will turn out on deeper analysis . . . to be fallacious."[38]

Orthodox scientists have also addressed basic epistemological issues. Thus, in the early 1970s, the AOJS Students' Questions Panel summarized "the standard orthodox approach" to evolution as follows:

> As one of our leading members has said, one of the bonuses of being an orthodox Jewish scientist is that one can become a better scientist, being used to scrutinise with the utmost care statements which others tend to take on trust. And it is not difficult to criticise the theory of evolution to show up its difficulties, its weaknesses, its speculative nature, its circular reasoning. The so-called facts of evolution are, it is said, not facts at all but extrapolations from fragmentary data backwards in time to a dim and unknown past.[39]

36. For an early attempt at an organization of Orthodox academics, see Ira Robinson, "Cyrus Adler, Bernard Revel, and the Prehistory of Organized Jewish Scholarship in the United States," *American Jewish History* 69 (1980): 497–505.
37. Michael N. Dobkowski, *Jewish American Voluntary Organizations* (New York: Greenwood Press, 1986), 76. Cf. Judy Siegel-Itzkovich, "Orthodox Scientists Found Strength in Numbers," *Jerusalem Post*, 12 July 1998, 10.
38. Alvin Radkowsky, "Faith in an Age of Scientific Achievement," in *Viewpoints on Science and Judaism*, ed. Tina Levitan (New York: Board of Jewish Education, 1978), 89–93, on 90–91. This position was not so subtly questioned by the AOJS Students' Questions Panel (Carmell and Domb, *Challenge*, 257), being characterized as an "Akeyda [binding, a reference to the binding of Isaac in preparation for the sacrifice] of the intellect" that demands that the doubter must "surrender" his "rational faculty" or consider himself to have abandoned "the ranks of believing Jews."
39. This position was adopted by Orthodox scientist Leo Levi in his *Torah and Science: Their Interplay in the World Scheme* (New York: Association of Orthodox Jewish Scientists; Jerusalem: Feldheim, 5743 [1983]), 104, and was criticized in Alvin Radkowsky, "Miracles," in *Encounter: Essays on Torah and Modern Life*, ed. H. Chaim Schimmel and Aryeh Carmell (Jerusalem: Association of Orthodox Jewish Scientists/Feldheim, 1989), 42–74, on 63.

> Evolution is, after all, "only a theory," it is argued, and as such it can have no power to influence our belief in the literal interpretation of *Bereishit* [Genesis] and the traditional time-scale.[40]

Recognizing that not every Orthodox scientist has subscribed to these views, the Students' Questions Panel indicated that the above approach "is not the only possible one."[41] Nevertheless, the "weaknesses and limitations" strategy has been adopted by many Orthodox thinkers.

A related concern of many Orthodox thinkers is that science is constantly in flux and offers an incomplete understanding of reality, whereas the Torah is unchangeable and perfect. Thus Elliot Pines has stated that the Torah

> is written by G-d, and as such it is a complete description of reality. Science['s] . . . subject matter is . . . not reality but a man-made model of reality. . . . The Torah, though packaged in a finite form, is Reality in all its infinity. Science is a model of Reality, and as such, despite delusions of grandeur, is as finite as the brain of Man. No matter how far it progresses, even if perfected within its limitations, it reflects an approximation to an infinitesimal speck of reality . . . all objective conflicts between Torah and Science, arise from this intrinsic fact.[42]

Finally, some Orthodox critics have accused science of being unscientific because it is subjective and dogmatic.[43] These criticisms have been applied specifically to evolution, which is widely portrayed as deviating from the standards of "objective" science.[44] Orthodox critics have charged evolution with not being "rigorous science,"[45] and lacking a "well-formulated hypothesis." Likewise the theory has been characterized as "an example of unrestrained speculation."[46]

40. AOJS Students' Questions Panel, "Actual and Possible Attitudes to Evolution," 256.

41. Ibid.

42. Pines, "Torah, Reality, and the Scientific Model," 3; Anonymous, "The Jewish Prospective on Evolution." Slifkin holds a similar position: "The Torah is a perfect description of all existence because it is the root of all existence . . . the universe being a physical manifestation of the Torah." Nosson Slifkin, *The Science of Torah: The Reflection of the Torah in the Laws of Science, the Creation of the Universe, and the Development of Life* (Southfield, MI: Targum; Nanuet, NY: Feldheim, 2001), 73.

43. Lewis Berenson, "The Evolution of Life," in Levitan, *Viewpoints on Science and Judaism*, 9–12, on 10; Morris Goldman, "Evolution by Natural Selection," ibid., 49–54, on 53; Herman Branover, "Torah and Science: Basic Principles," in Schimmel and Carmell, *Encounter*, 232–42, on 236.

44. Nathan Aviezer, "Misreading the Fossils: The Dark Side of Evolutionary Biology," abstract of article, www.biu.ac.il/JH/BDD/engabs.htm. Cf. Moses L. Isaacs, "The Challenge of Science," in Jung, *The Jewish Library*, 173.

45. Radkowsky, "Judaism and the Atomic Age," 240.

46. Sol Roth, *Studies in Torah Judaism: Science and Religion* (New York: Yeshiva University, 1967), 54.

Schroeder claims that the theory of evolution in its current form is not a true scientific theory, but "merely a description of the 'punctuated' jumps in the fossil record."[47] He also attempts to discredit the theory on the grounds that "all calculations of probability say no to the assumption of randomness being the driving force behind life's development."[48]

Such technical critiques of evolution are read and digested by rabbis, who disseminate these views to Orthodox audiences in their sermons and publications. A good example of a popular, rhetorical treatment of these themes, informed by a reading of some of the Orthodox scientists mentioned above, is that of Rabbi Avigdor Miller, prolific author, public speaker, and past director of a yeshiva in Flatbush:

> We see the *yad Hashem* [hand of God] in nature. . . . In a book of molecular biology, there are six thousand entries in the index . . . two entries on evolution, and the writer said in his preface, one of the purposes of biology is to teach people the principle of evolution, and in the entire book nothing is mentioned. Two places! In these two places it doesn't say any proof for evolution.[49] It just said it evolved. How could it evolve? It's so complicated, and if one of the elements is missing, and there are hundreds of elements, precise arrangements that had to be mathematically exactly correct.[50]

INTEGRATION OF SCIENCE AND TORAH

In contrast to those Orthodox Jewish scientists and rabbis who oppose evolution, a few have adopted integrationist strategies. While they do not closely follow the medieval rationalist project as exemplified by Saadia, they do subscribe to the medieval notion that the Torah can be integrated with science.

One example is Judah Landa, whose *Torah and Science*[51] contains a sustained polemic against Orthodox Jews who "motivated by considerations other than science . . . persist in a stubborn refusal to accept the tower of [scientific]

47. Gerald Schroeder, *The Hidden Face of God: How Science Reveals the Ultimate Truth* (New York: Free Press, 2001), 91. On Schroeder see Cherry (chapter 7) in this volume.
48. Schroeder, *Hidden Face of God*, 100; cf. 120.
49. Miller seems to have derived this criticism from Gottlieb, "The Theory of Evolution"; however, while Miller says evolution is referenced twice in the index to the book, Gottlieb claims there are four references to evolution.
50. Avigdor Miller, "Diamonds on the Road" (audiotape E-235) (Brooklyn: Yeshiva Gedolah Bais Yisrael, 2002). Miller became an important influence on the thought of Sara Yoheved Riegler, "Confessions of a Creationist."
51. Landa, *Torah and Science*. On Landa, see Cherry (chapter 7) in this volume.

evidence."[52] He argues that scientific research is an activity suitable for Orthodox Jews and that its conclusions cannot be contradicted by Torah. According to Landa, those Orthodox fundamentalists who oppose science are mistaken in their adamant refusal to admit that traditional rabbinic interpretations could possibly be wrong. In adopting this attitude they continue "to search for baseless objections to the powerful evidence."[53] By contrast, Landa accepts the findings of science and assumes that if the medieval rabbis were alive today, they "would see fit to reconcile their interpretation of the six days and the entire story of creation with the evidence."[54] In a somewhat utopian conclusion, he looks forward to the convening of

> a conclave of prominent and open-minded Orthodox rabbis for the purpose of considering the adoption of the following resolution:
>
>> Be it made known that neither the scientifically established age of the earth . . . nor the general outline of the theory of evolution is in conflict with the Torah. The story of creation in Genesis and the entire Torah can readily be interpreted in such a manner that there is no conflict with any established principle of science. None of the fundamental tenets of Judaism regarding the existence of God, creation, divine intervention in the affairs of mankind, and the occurrence of miracles contradicts any established principle or theory of science. Any representations to the contrary, made in the past by people and organizations in the name of Judaism or by members of the scientific community, are erroneous and based on misunderstandings. We regret the confusion and misconceptions which have been propagated in the wake of our silence on this important issue for so long a time.[55]

Less radical in tone, but equally opposed to those Orthodox Jews who reject the scientific consensus on the theory of evolution is the electrical engineer and physicist Baruch Sterman, who decries the "lack of willingness within the Jewish intellectual community to face Darwinism in an open-minded fashion"[56] and criticizes the antievolutionist views of such prominent Orthodox scientists as electrical engineer Leo Levi and physicists Herman Branover and Nathan Aviezer.

Finally, in an article published in 1990, Yeshiva University biology professor Carl Feit agreed that evolution is "central to the whole enterprise of biology

52. Landa, *Torah and Science*, 273.
53. Ibid., 326.
54. Ibid., 325.
55. Ibid., 349.
56. Sterman, "Judaism and Darwinian Evolution," 62.

today," having withstood "one hundred years of the most intense analysis." There is, he wrote, "no alternative . . . theory to explain the phenomena with which it deals."[57] Feit has adopted a nonfundamentalist, nonliteral interpretation of Torah,[58] similar to that formulated by medieval scholars such as Maimonides. The Torah, he claims, may "allow for the existence of several, even mutually exclusive truths," just as halakhah (Jewish law) allows for the notion of multiple truths, each truth being judged according to its own criteria. Not wishing to paint himself entirely in a corner, however, he nonetheless does not believe that Torah's truth can be totally open to relativization.[59]

TRANSCENDENCE OF SCIENCE

The majority of Orthodox Jews who have sought to avoid a direct confrontation with science have not, however, pursued an integrationist strategy. Rather, they have sought to transcend the problem by using concepts derived from rabbinic aggadah (nonlegal arguments) in general and kabbalah in particular.[60]

This stance can be readily seen in the argument of Susan Schneider, who is associated with the Torah Science Foundation:

> Tradition teaches that the entire creation chapter did actually happen and in a physical sense, but on an entirely different level than what we now know as the physical plane.[61]

Pursuing a similar line of thought, Rabbi Dovid Brown has written:

> Our point . . . is not, however, to refute the theory of evolution in order to justify our belief in the divine creation of the universe. As the descendants of those who stood at Har [Mount] Sinai and accepted the Torah, we do not need the assent of secular intellectuals to maintain our faith. . . . However *Chazal* [the ancient rabbinic authorities] tell us . . . "Falsehood cannot exist without some admixture of truth." What element of truth is there in this falsehood?[62]

Similarly, Elliot Pines, drawing on kabbalistic concepts, asserts that "our universe is a simulation" and concludes triumphantly: "Science can, in all

57. Carl Feit, "Darwin and Drash: The Interplay of Torah and Biology," *Torah u-Madda Journal* 2 (1990): 25–36, on 29–30.
58. Ibid., 31. Sol Roth asserts that "the biblical chapter on creation has correctly been assigned a metaphorical interpretation by many thinkers who adhere without compromise to Jewish tradition"; see Roth, *Studies in Torah Judaism*, 53.
59. Feit, "Darwin and Drash," 28–29.
60. Ira Robinson, "Kabbala and Orthodoxy: Some Twentieth Century Interpretations," unpublished paper presented at the American Academy of Religion, 1987.
61. Schneider, "Evolutionary Creationism," 7.
62. Brown, *Mysteries of the Creation*, 276.

sincerity, demonstrate a fact of the physical world that seems to contradict Torah. However science is, in actuality, modeling what is in itself merely the present state of a simulation."[63]

Rabbi Abraham Isaac ha-Kohen Kook is well known for having expressed the view that evolution possesses a greater affinity with the secret teachings of kabbalah than with all other philosophies.[64] A contemporary rabbi, Judah Yudel Rosenberg, who emigrated from Poland to Canada, wrote that Darwin's theory of evolution possesses a certain distorted sense of the true revelation of the Zohar, the main source of kabbalistic doctrine, which asserts that under our earth there exist seven lands, all of which are inhabited by men not of the seed of Adam. He speculated that Darwin had been aware of the Zohar's view about the creatures of the lands "down under" and had established his "mad heresy" that men developed from smaller animals, like monkeys, and they, in turn, had developed from still smaller creatures. Actually, states Rosenberg, "[i]n several places in the *Zohar* the opposite is stated that the apes are the descendants of sinful men. Something similar is agreed upon by the honest scientists of the nations of the world."[65]

A number of popularizers of kabbalah, including Rabbis Judah Ashlag, Yehuda Brandwein, Philip S. Berg, Aryeh Kaplan,[66] and David Sheinkin, have also attempted to link kabbalah and science.[67] In particular, Sheinkin, a psychiatrist who has written extensively on mysticism, has stated that kabbalah is consistent with belief in some form of evolutionary process.[68]

In recent decades, this position has been adopted by many Orthodox Jewish scientists, not only rabbis. For example, Alvin Radkowsky has argued that "the increasingly remarkable progress of mankind in the last four

63. Pines, "Torah, Reality, and the Scientific Model."

64. Abraham Isaac ha-Kohen Kook, *Lights of Penitence*, trans. Ben Zion Bokser (New York: Paulist, 1978), 220, 306. See also Feit (chapter 9) in this volume. A number of contemporary Orthodox thinkers on evolution cite Kook, e.g., Cyril Domb, "Biology and Ethics," *Proceedings of the Association of Orthodox Jewish Scientists* 3–4 (1976): 9–19, on 15–16.

65. Ira Robinson, "Kabbalist and Communal Leader: Rabbi Yudel Rosenberg and the Canadian Jewish Community," *Canadian Jewish Studies* 1 (1993): 41–58; Judah Yudel Rosenberg, *Zohar Torah* (New York: n.p., 1924), Genesis, 62; Rosenberg, *Ha-Zohar ha-Kadosh* (Bilgoraj, Poland: Wajnberg, 1929), Psalms, 23. Rosenberg's opinion is echoed by Mosheh Epstein, *Torah Verified by Science* (New York: M. Epstein, 1928), 17. This opinion parallels that of the Christian creationist, George McCready Price; see Numbers, *The Creationists*, 85.

66. Of Kaplan it was said, "The Midrash could illuminate the laws of relativity, and developments in biogenetics could explain Messianic prophecies." Y. Elkin in the preface to Aryeh Kaplan, *Facets and Faces* (Jerusalem: Moznaim, 1993), 10.

67. Robinson, "Kabbala and Orthodoxy."

68. David Sheinkin, *The Path of the Kabbalah* (New York: Paragon, 1986), 148.

centuries was actually prophesied in the kabbalah."[69] Likewise Eliezer Zeiger of the Torah Science Foundation writes:

> We thus see that the inner wisdom of the Torah clearly reveals the operation of evolutionary processes in creation. . . . However, the Holy Izhbetzer [Rabbi Mordecai Joseph Leiner (1802–54)],[70] the author of Mei Sheloah, published about the same time than [*sic*] Darwin's Origin of Species, wrote in Pei [sic] Breshit:
>
>> In the beginning, G-d created all the creations. Then the creations understood their limitation that they did not have anybody that would unite their life with the Holy One, and that by means of man all the sayings of the world will be united with the Creator, and that the inanimate gives its power to the plants and the plants give their power to the animals and the animals give their power to the human *hamedaber* [the one who can talk], so that man will worship with his power the Holy One. When the creations saw what they were lacking, they use their power to create a *Hitarerutah le maala* [an awaking above] for the creation of man. And Elokim [God] said Let us make man, and the Holy One told the creations that all of them give of their power to contribute to the creation of man, so that man will have a part of all of them, so that if man will be in need, they will all help him because when it is bad for man it is bad to all creatures like in the generation of the flood, and when is good for man is good for all creatures as well.
>
> How do we relate to the Izhbetzer's teaching? As a metaphor? This is a very important methodological question. Many people erroneously think that the Torah is only a moral discourse. . . . However, a thorough study of the inner wisdom of the Torah readily reveals its outstanding precision. So if we can relate to the Izhbetzer's teaching as a precise statement, we realize that it has stunning biological implications. It hints that the sharing of many biological features by man, animals and plants arose from a contribution from all created organisms to the formation of man, in order for them to be their partner in the praising of G-d.
>
> We thus see that an analysis of Genesis based on the inner wisdom of the Torah unifies the Torah and science views of the origin of the world and of life. It shows that the creative forces that shape all creatures of the world are Divine forces, and that the unfolding of these major creations are micro-evolutionary forces that follow natural laws and that are amenable to scientific analysis.[71]

69. Radkowsky, "Miracles," 69.
70. Morris M. Faierstein, *All Is in the Hands of Heaven: The Teachings of Rabbi Mordecai Joseph Leiner of Izbica* (New York and Hoboken: Yeshiva University Press, 1989).
71. Zeiger, "Kosher Evolution."

The controversial "zoo rabbi," Nosson Slifkin, likewise grounds his model of evolution, as of Torah and the universe in general, on the kabbalistically tinged Hebrew term *hishtalshelut*—"the sense of the gradual unfolding of fundamental patterns from simple unity to complex multiplicity."[72]

To Whom It May Concern: The Evolution of an Argument

In 1978 the AOJS Students' Questions Panel expressed the conviction that "the conflict between 'religion' and 'evolution' has outlived its usefulness and it was high time it was allowed a quiet demise."[73] Nonetheless, at the beginning of the twenty-first century, the argument among people who consider themselves Orthodox Jews over the validity of the theory of evolution continues with no sign of closure. All the strategies for dealing with the issue described above have contemporary advocates. In large measure, this is due to a combination of the diversity of opinion among Orthodox Jews with respect to the validity of "secular" science with the absence of a generally accepted process within Orthodox Jewry to resolve ideological or doctrinal issues. It is, however, possible to identify some significant changes in attitudes towards evolution during the last half of the twentieth century.

First of all, one can detect a shift of opinion away from a focus on scientific arguments against the theory of evolution. In a 1998 article about the AOJS, journalist Judy Siegel-Itzkovich quoted the founder of AOJS, Elmer Offenbacher, who stated that "in the old days, the matter of evolution vs. religion was a hot topic; some people were obsessed by it." He added that religious scientists had gradually accepted that there was no real conflict between the two.[74] This perceived relative decline in the intensity of the issue among Orthodox Jewish scientists parallels the decline in the AOJS itself, from a membership of nearly two thousand in the early 1960s to about eight hundred in the late 1990s. In part this decline reflects the security felt by the increasing number of Orthodox Jews who pursue scientific careers, while at the same time assigning their scientific and religious activities to separate domains.[75]

72. Slifkin, *Science of Torah*, 215; cf. 73, 89, 98, 164.

73. AOJS Students' Questions Panel, "Actual and Possible Attitudes to Evolution," 268.

74. Siegel-Itzkovich, "Orthodox Scientists." This decline of interest in evolution among Orthodox scientists might possibly be reflected in Aryeh Carmell's introduction to Slifkin's *Science of Torah*, in which he stated that Orthodox Jewish scientists should be in the forefront of these debates.

75. Robert Werman, a contemporary biologist and Orthodox Jew, expressed his personal position thus: "As a Professor in a Biology Institute (at the Hebrew University) I teach evolution; as a religious Jew I daven and associate with Haredim [ultra-Orthodox]." Mail.Jewish, 10 September 1993, http://www.ottmall.com/mj_ht_arch/v9/mj_v9i16.html#CCC.

Another change is the significant growth of interest in the "transcendence of science" strategy grounded in kabbala. One key factor is the increasing level of "ḥaredization"—the tendency towards ultra-Orthodox ideologies—within contemporary Orthodox Jewry.[76] Most Ḥaredim tend to ignore science, but those who do take it seriously often opt for a mystical reconciliation of Torah and science. Moreover, in recent decades, kabbalistic explanations of the world have become ubiquitous, enthusiastically accepted by many Jews and non-Jews alike.[77] In this context, the popularity of the strategy of transcendence of scientific evolutionary theory by kabbalistic concepts is not at all surprising and brings to mind the statement of Isaac Bashevis Singer:

> When I read that a stone consisted of trillions of molecules, constantly in motion and that these molecules consisted of atoms, and that these atoms were in themselves complicated systems, whirls of energy, I said to myself, "That's the cabala after all."[78]

Finally, it is important to place Orthodox discussions of evolution in the context of outreach activities aimed at non-Orthodox Jews, which has been an important dynamic element within the Orthodox community.[79] For those firmly within the ultra-Orthodox camp, where any conflict between Torah and science is decided in favor of Torah, arguments about evolution are not that important. Similarly, arguments about evolution are not important for those Orthodox Jews who accept modern science and reinterpret Torah to various degrees to accommodate scientific results. Rather, the audience for those refutations and explanations of evolution detailed in this article consists principally of *ba'alei teshuvah* (Jews from non-Orthodox backgrounds who have recently become Orthodox)[80] and those who are seeking to draw non-Orthodox Jews into the Orthodox fold.[81] Thus, many of the publications cited in this paper are

76. On contemporary Orthodoxy see Haym Soloveitchik, "Rupture and Reconstruction: The Transformation of Contemporary Orthodoxy," *Tradition* 28, no. 4 (1994): 64–130.

77. Howard Gontovnick, "Kabbalah 2000: The Emergence of a New Religious Movement," presented at the Society for the Scientific Study of Religion, 2003.

78. Isaac Bashevis Singer, *Love and Exile* (New York: Doubleday, 1984), 16.

79. On this phenomenon, see M. Herbert Danziger, *Returning to Tradition: the Contemporary Revival of Orthodox Judaism* (New Haven: Yale University Press, 1989); Janet O'Dea Aviad, *Return to Judaism: Religious Revival in Israel* (Chicago: University of Chicago Press, 1983); Lynn Davidman, *Tradition in a Rootless World: Women Turn to Orthodox Judaism* (Los Angeles: University of California Press, 1991).

80. See Aaron J. Tapper, "The 'Cult' of Aish Hatorah: Ba'alei Teshuvah and the New Religious Movement Phenomenon," *Jewish Journal of Sociology* 44 (2002): 5–25; Danziger, *Returning to Tradition*, 282–87.

81. Cf. Golani, review Spetner, which recommends the works of Orthodox Jewish science mainly to "older students and adults who must contend with the mindset of the secular world." See also Goldman, "Evolution by Natural Selection," 50.

aimed at *ba'alei teshuvah* or potential *ba'alei teshuvah* who might find Orthodoxy more attractive if the seeming contradictions between their conception of Judaic belief and scientific theory could be resolved. As Slifkin states when discussing the age of the universe, "[T]he scientific evidence for an old universe is . . . so vast and overwhelming that it is rather unwise to simply wave it away (and the effects on Jewish outreach efforts are disastrous)."[82] Thus, ultimately, the issue of the relationship between Torah and the theory of evolution within Orthodox Judaism is connected to the way Orthodoxy is presented to those Jews who are not born Orthodox but are in the process of coming to accept the ideas and ideology of Orthodox Judaism.

82. Slifkin, *Science of Torah*, 90.

PART TWO

Social Uses of Evolution: Anti-Semitism, Racism, and Zionism

The second focus of this volume is the deployment of evolutionary ideas, by both Jews and others, in the domains of race, anti-Semitism, and Zionism. At the center of this cluster of issues lies the recurrent use over the last century and a half of evolutionary ideas to characterize Jews. This leads to two large historical questions, which are addressed by several works cited in the list of recommended readings. First, how have others, often anti-Semites, deployed evolution in order to portray Jews as different and ultimately inferior? This question requires an analysis of the relationship between scientific theory, scientific practice, and the social and political agendas of those who developed and implemented anti-Semitic philosophies. Second, how did Jews use evolutionary ideas to define Jewish identity and pursue their own social analysis and communal programs? Evolution has not only been a tool in the anti-Semite's armory; Jewish and pro-Semitic thinkers have used it in a variety of other ways and in various historical contexts.

Before presenting the papers in this section, it should be noted that notions of race and racism have a complex history that long predates Darwin's writings. Based on the reports of travelers, eighteenth-century naturalists and philosophers exhibited their preoccupation with classification by characterizing different peoples by their outward appearance, which was often seen as reflecting their intellectual and moral qualities. While Rousseau enthused about "the savage," the more typical response was to emphasize the vast differences between primitive peoples and the Enlightened Europeans with their developed intellects, civilized societies, and refined manners. Yet with the emergence of ethnology in the early nineteenth century, long-established assumptions about human history were challenged as ethnologists sought to build their science on empirical evidence. Notions of race also became increasingly hardened, not only specifying the races as intrinsically separate and different, but also in many cases establishing a hierarchy of races headed by the European.

Racial ideas developed in many areas, including German romantic historicism, which emphasized the individuality of cultures and peoples, and philology, which developed the theory of an original Indo-European language that had been imported from Asia into Europe by the migration of "Aryan" peoples. As the nineteenth century progressed, writers began to combine these various developments, one of the most systematic being the Comte de Gobineau, who offered in his *Essai sur l'Inégalité des Races Humaines* (1853–55) a comprehensive social and political analysis based on the view that the depravity of the age had resulted from racial mixing. For Gobineau, the white or Aryan race possessed supreme human virtues, which could be preserved only through racial purity. However, the Aryan purity of the German people was threatened by interbreeding with degenerate, impure, mixed races such as Latin and Semitic peoples, including Jews.

Particularly in Britain and America a central issue of the period was whether all human races have a common origin (monogeny) or are separate and essentially distinct, like different species (polygeny). Like many other monogenists James Cowles Prichard defended the Mosaic view that all races were descended from Adam and Eve. Beginning with his *Researches into the Physical History of Man* (1813) Prichard sought to explain human differences by appealing to contingent factors, such as climate, while also acknowledging the importance of culture and inheritance. On the other side of the argument Robert Knox characterized each race by its intrinsic physical features and mental characteristics in his controversial *Races of Man* (1850). For Knox, Jews constituted a separate race that he described in familiar and derogatory terms, pointing, for example, to the Jew's "large, massive, club-shaped nose." He also aligned Jews with Africans and considered that Jews, like Africans, were an alien race that could never become true Europeans but threatened European society through assimilation.[1]

Although Darwin avoided addressing the evolution of humankind in the *Origin of Species*, which was published a few years after Gobineau's and Knox's books, his implicit views were apparent to many readers and were made explicit in *The Descent of Man* (1871). In some respects Darwin's views represented a break with both monogenist and polygenist theories, since he

1. Robert Knox, *The Races of Man: A Fragment* (London: Renshaw, 1850), 51, quoted in John M. Efron, *Defenders of the Race: Jewish Doctors and Race Science in Fin-de-Siècle Europe* (New Haven: Yale University Press, 1994), 49–55.

portrayed the varieties of humankind as having evolved from the higher animals. But in other ways he was closer to the monogenists, since the different races were not distinct but were historically interrelated. Yet in characterizing the continuity between humans and other animal species Darwin deployed traditional hierarchical notions in placing the "savages" closest to the primates and the civilized Europeans at the top. Moreover, he conceived an overall historical progress—mental, moral, and to some extent biological—from the savage to civilized state. While acknowledging the superiority of the European races, he also on occasions conceived the various races as locked in a struggle with each other.[2] Although Darwin did not think of Jews as constituting a separate race, later writers, especially in Germany, sought not only to apply Darwinian principles to society but to identify Jews as constituting an inferior race that was locked in the struggle for existence with the Aryan race.

While it has often been claimed that the theory of evolution provided a scientific underpinning for German racial theory and anti-Semitism, in chapter 4 Richard Weikart analyzes, in a detailed and sensitive manner, the roles played by Darwinian evolutionary theory in the construction of anti-Semitic ideology. Chapters 5 and 6, by Paul Weindling and Raphael Falk, explore the relationship between evolutionary ideas and the political agendas of Jewish scientists and intellectuals. By examining the career of Ignaz Zollschan (1877–1948), one of the leading advocates for a Jewish race science, Weindling shows how his scientific views changed in response to the threat posed by Nazi race theory. Zollschan also emerges as a political activist who helped orchestrate international opposition to Nazi ideology. Finally, in the sixth chapter, Falk demonstrates how Zionists used evolutionary ideas in their ideologies and in their practical programs, both before and after the founding of the State of Israel.

These essays raise important questions about the relationship between scientific theorizing and both political ideologies and social practices. With the sizable literature on racial theory in the German-speaking world a clearer understanding is emerging of the often complex processes by which evolutionary ideas—bolstered by their social authority—were used to justify racial discrimination, dehumanization, and ultimately genocide. While not undervaluing the research on Germany, it is also important to investigate further how evolutionary ideas were used to underpin anti-Semitism in other countries, including America, France, and Russia, not only in the crucial interwar

2. Nancy Stepan, *The Idea of Race in Science: Great Britain, 1800–1960* (Basingstoke: Macmillan, 1982), 47–82.

period but also subsequently. More specifically, building on the work of Weindling, Efron, Hart, Müller-Hill, and others, a better understanding needs to be gained of how both Jewish and non-Jewish biologists, doctors, and psychologists positioned themselves at a time when the theory of evolution was being deployed by many leading scientific authorities for patently anti-Semitic purposes. Moreover, building on Falk's innovative paper, further research needs to be pursued on how some of the same scientific resources were deployed by Zionist thinkers in their visions for the future of Jewry.[3]

3. One interesting example, suggested by one of the referees, is Aḥad Ha-Am (pen name of Asher Ginsberg, 1856–1927), the leading philosopher of cultural Zionism. While rejecting Herzl's vision of a Jewish homeland, Aḥad Ha-Am advocated the building a Jewish cultural center in Palestine, to serve as a center of identity for world Jewry, a vision he supported by interpreting world and Jewish history using social Darwinist ideas. See Aḥad Ha-Am, *Selected Essays*, trans. Leon Simon (Philadelphia: Jewish Publication Society of America, 1944); Steven J. Zipperstein, *Elusive Prophet: Ahad Ha'am and the Origins of Zionism* (Berkeley: University of California Press, 1993), 93.

4

The Impact of Social Darwinism on Anti-Semitic Ideology in Germany and Austria, 1860–1945

Richard Weikart

Many scholars have explained how social Darwinism underpinned Hitler's militarism and racism by showing that Hitler's doctrine of racial struggle and his policy of racial extermination were largely shaped by social Darwinist racial thought, which was prominent among early twentieth-century German scientists and physicians.[1] In their analyses of the period historians have generally defined "social Darwinism" as the program to apply Darwinian principles to human society, while insisting that humans are subject to an inexorable struggle for existence. Indeed, social Darwinists exulted in the beneficence arising from human rivalry, even when it resulted in the death of the losers, since it would promote biological progress for the human species. By the term "social Darwinist racism" is meant the struggle for existence between human races that yields biological advance, while ultimately resulting in the extermination of the "inferior" races. This does not mean that all social Darwinist racists *promoted* the extermination of inferior races, although most conceived of the European imperialists as biologically, culturally, and intellectually superior to the aboriginal peoples being decimated by them. We should also remember that not all Darwinists embraced social Darwinism or racism, since neither was a necessary concomitant of the scientific theory of evolution. However, like the vast majority of their European contemporaries, most leading Darwinists in the late nineteenth century maintained ideas of inequality between the races and believed that racial struggle resulting in the extermination of some races played an integral role in the progressive selectionist process leading to biological improvement.[2]

1. For a full discussion of the historiography, see Richard Weikart, *From Darwin to Hitler: Evolutionary Ethics, Eugenics, and Racism in Germany* (New York: Palgrave Macmillan, 2004). Two examples among many are Ian Kershaw, *Hitler*, 2 vols. (New York: Norton, 1998–2000); Eberhard Jäckel, *Hitler's Weltanschauung: A Blueprint for Power*, trans. Herbert Arnold (Middleton, CN: Wesleyan University Press, 1972).

2. The pioneering work on social Darwinism was Richard Hofstadter, *Social Darwinism in American Thought*, rev. ed. (Boston: Beacon Press, 1955); Robert C. Bannister disputed Hofstadter's claim that social Darwinism was influential in *Social Darwinism: Science and Myth in Anglo-American Social Thought* (Philadelphia: Temple University Press, 1979). The role of

The social Darwinist influence on Hitler's—and other Germans'—anti-Semitism is not straightforward, since Darwinism and anti-Semitism are not necessarily connected. No necessary inference can be drawn from Darwinism to the status of Jews, and some Darwinian biologists, such as Arnold Dodel, a botanist at the University of Zurich in the late nineteenth century, staunchly opposed anti-Semitism on biological grounds.[3] Many Jewish scientists and thinkers, especially those with secular leanings, adopted Darwinism with alacrity. The prominent Zionist author Max Nordau, for example, considered Darwinism an integral component of his scientific worldview, and he constructed an ethical philosophy based on Darwinian theory.[4]

Although most Darwinian biologists in nineteenth-century Germany wrote little or nothing about Jews, many stressed racial competition as a form of the human struggle for existence. Some clearly approved the disappearance of "inferior races" before the European advance.[5] When discussing the extinction through competition of inferior groups, most social Darwinist racists in the late nineteenth century included American Indians, Australian aborigines, black Africans, and East Asians. However, many anti-Semitic racists in late nineteenth- and early twentieth-century Germany also encompassed social Darwinist racism and integrated it into their racial ideology. This allowed them to claim scientific status for their view that unequal races were locked in an eternal struggle for existence. Moreover, in the eyes of many of these German anti-Semites, Jews posed the greatest and most immediate threat in this competition to the death. By fostering this fixation on racial competition and even racial extermination, social Darwinist racism helped transform earlier forms of anti-Semitism into an ideology of annihilation.

---

social Darwinism in Europe and the U.S. is currently in dispute. An excellent recent analysis of social Darwinism is Mike Hawkins, *Social Darwinism in European and American Thought, 1860–1945: Nature as Model and Nature as Threat* (Cambridge: Cambridge University Press, 1997); see also Paul Weindling, *Health, Race, and German Politics between National Unification and Nazism, 1870–1945* (Cambridge: Cambridge University Press, 1989), ch. 1; Richard Weikart, "The Origins of Social Darwinism in Germany, 1859–1895," *Journal of the History of Ideas* 54 (1993): 469–88.

3. Arnold Dodel, *Moses oder Darwin? Eine Schulfrage* (Zurich: Schmidt, 1889), 98; Arnold Dodel to Bertha von Suttner, 16 April 1890, Suttner-Fried Collection, League of Nations Archives, United Nations Library, Geneva.

4. Max Nordau, *Morals and the Evolution of Man*, trans. Marie A. Lewenz (trans. of *Biologie und Ethik)* (London: Cassell, 1922).

5. Richard Weikart, "Progress through Racial Extermination: Social Darwinism, Eugenics, and Pacifism in Germany, 1860–1918," *German Studies Review* 26 (2003): 273–94.

Jews in Germany and elsewhere in Europe endured centuries of hostility and persecution from the Christian majority—including stereotyping, discrimination, and pogroms. Yet Christian anti-Semites generally accorded Jews a limited amount of toleration; usually their goal was conversion, which would give Jews the same social and legal status as Christians. Many scholars have noted the late nineteenth century shift from traditional forms of Christian anti-Semitism to secular racial anti-Semitism.[6] Although the new racial anti-Semitism of the nineteenth century retained many of the long-standing Jewish stereotypes, such as their alleged immorality, it closed the door to assimilation, since Jews could not discard their immoral character, which was now grounded in their biological essence. The only solution was to get rid of the Jews. The interaction between Darwinian theory and anti-Semitic thought in this shift toward biological racism has not been adequately explored.

While many recent works on Nazi racism have stressed the importance of scientific racism, some scholars have argued that mystical, idealist, and irrationalist racial theories were more influential in shaping Nazi anti-Semitism.[7] In the ensuing discussion, I do not deny the contribution of

6. Jacob Katz, *From Prejudice to Destruction: Anti-Semitism, 1700–1933* (Cambridge: Harvard University Press, 1980); Paul Massing, *Rehearsal for Destruction: A Study of Political Anti-Semitism in Imperial Germany* (New York: Harper, 1949); George L. Mosse, *Toward the Final Solution: A History of European Racism* (New York: Fertig, 1978); Leon Poliakov, *The Aryan Myth: A History of Racist and Nationalist Ideas in Europe* (London: Sussex University Press, 1974); Peter Pulzer, *The Rise of Political Anti-Semitism in Germany and Austria* (New York: Wiley, 1964). I follow many historians in using the term "anti-Semitism" to refer to all forms of persecution or discrimination against Jews, whether based on religion, culture, or race (even if it occurred before the term "anti-Semitism" was coined in the 1870s).

7. On scientific racism, see, among others, Michael Burleigh and Wolfgang Wippermann, *The Racial State: Germany, 1933–1945* (Cambridge: Cambridge University Press, 1991); Benoit Massin, "The 'Science of Race,' " in *Deadly Medicine: Creating the Master Race*, ed. Susan D. Bachrach and Dieter Kuntz (Washington: United States Holocaust Memorial Museum, 2004); Henry Friedlander, *The Origins of Nazi Genocide: From Euthanasia to the Final Solution* (Chapel Hill: University of North Carolina Press, 1995); Benno Müller-Hill, *Murderous Science: Elimination by Scientific Selection of Jews, Gypsies, and Others, Germany, 1933–1945*, trans. George R. Fraser (Oxford: Oxford University Press, 1988); Robert Proctor, *Racial Hygiene: Medicine under the Nazis* (Cambridge: Harvard University Press, 1988). Works emphasizing mystical or irrationalist racism include George Mosse, *The Crisis of German Ideology: Intellectual Origins of the Third Reich* (New York: Grosset and Dunlap, 1964); Mosse, *Toward the Final Solution*; Saul Friedländer, *Nazi Germany and the Jews*, vol. 1, *The Years of Persecution, 1933–1939* (New York: HarperCollins, 1997); Fritz Stern, *The Politics of Cultural Despair: A Study in the Rise of the Germanic Ideology* (Garden City, NY: Anchor Books, 1965). Klaus Fischer, *The History of an Obsession: German Judeophobia and the Holocaust* (New York: Continuum, 1998), tends to avoid

mystical forms of anti-Semitism to Nazi ideology. The mystical racist Paul de Lagarde, for instance, owed little or no intellectual debt to Darwin, but his mid-nineteenth-century racial ideas contributed important elements to early twentieth-century anti-Semitic discourse and were highly regarded by Nazis. Another example is the anti-Semitic writer Julius Langbehn, who adopted an antiscientific and anti-Darwinian position in his incredibly popular *Rembrandt als Erzieher* (1890, Rembrandt as Educator).[8] However, as I explain later, idealist and mystical anti-Semitic thinkers, such as Jörg Lanz von Liebenfels, Houston Stewart Chamberlain, and Ludwig Schemann, often incorporated elements of social Darwinism into their racial theories. Some historians have overemphasized non-evolutionary factors: for example, Saul Friedländer, while crediting Chamberlain and Schemann with helping develop "redemptive anti-Semitism," ignores the scientific side of their racial ideologies and emphasizes their debt to mystical anti-Semitism.[9]

Darwin, Haeckel, and Büchner on Race

Darwin himself believed that human races were unequal and were locked in a human struggle for existence. Even though he opposed slavery and at times expressed sympathy for those of other races in his Beagle journal, he also expressed the view that the "varieties of man act on each other, in the same way as different species of animals—the stronger always extirpate the weaker."[10] Later in life he wrote to a colleague that the "more civilised so-called Caucasian races have beaten the Turkish hollow in the struggle for existence. Looking to the world at no very distant date, what an endless number of the lower races will have been eliminated by the higher civilised races throughout the world."[11] He articulated this same principle in *The Descent of*

the term "mystical anti-Semitism" and emphasizes the integration of social Darwinism and eugenics into anti-Semitism in Germany, as does Richard J. Evans, *The Coming of the Third Reich* (New York: Penguin, 2004).

8. Julius Langbehn, *Rembrandt als Erzieher*, 43rd ed. (Leipzig: Hirschfeld, 1893). Both Lagarde and Langbehn are discussed at length in Stern, *Politics of Cultural Despair*.

9. Friedländer, *Nazi Germany*, ch. 3.

10. Quoted in Barry W. Butcher, "Darwinism, Social Darwinism, and the Australian Aborigines: A Reevaluation," in *Darwin's Laboratory: Evolutionary Theory and Natural History in the Pacific*, ed. Roy MacLeod and Philip F. Rehbock (Honolulu: University of Hawaii Press, 1994), 380. Butcher shows the influence of ideas about racial extermination on the formulation of Darwin's views on human evolution.

11. Francis Darwin, *Charles Darwin: His Life Told in an Autobiographical Chapter, and in a Selected Series of His Published Letters* (London: Murray, 1902), 64.

Man, claiming, "At some future period, not very distant as measured by centuries, the civilised races of man will almost certainly exterminate and replace throughout the world the savage races."[12] Darwin was not a militarist, and he often expressed sympathy for those of "lower" races, so he would undoubtedly have been horrified by those coming after him who espoused racial extermination, but nonetheless, he did consider racial extermination an inescapable natural process that contributed to human evolution.

Only rarely did Darwin mention Jews in his writings, even in his discussions of race in *The Descent of Man*. Never did he exhibit the slightest anti-Semitic inclination in those passages, and once he even remarked that "Europeans differ but little from Jews."[13] Darwin's racism was directed primarily at "barbarians," as he called them, such as the Fuegians, Australian aborigines, American Indians, and black Africans. It was not aimed at the civilized Jews.

The most influential Darwinian biologist in Germany, Ernst Haeckel, emphasized racial inequality even more than Darwin did. He not only claimed that humans should be divided into twelve separate species, but he even divided those twelve species into four genera. In the first edition of his popular book, *Natürliche Schöpfungsgeschichte* (1868, The Natural History of Creation), he alleged that "the differences between the lowest humans and the highest apes are smaller than the differences between the lowest and the highest humans."[14] This idea was restated numerous times in his popular writings, sometimes even more crudely.

What role did Jews play in Haeckel's racial scheme? First, only rarely did Haeckel overtly mention Jews in his discussions of races. However, the racial

12. Charles Darwin, *Descent of Man*, 2 vols. (Princeton: Princeton University Press, 1981), 1:201. Most scholars agree that racial struggle is an integral part of Darwin's account of human evolution, and some even explicitly discuss the role of racial extermination in his theory—see Adrian Desmond and James Moore, *Darwin* (New York: Joseph, 1991), pp. xxi, 191, 266–68, 521, 653; Robert M. Young, "Darwinism *Is* Social," in *The Darwinian Heritage*, ed. David Kohn (Princeton: Princeton University Press, 1985), 609–38; John C. Greene, "Darwin as Social Evolutionist," in *Science, Ideology, and World View: Essays in the History of Evolutionary Ideas* (Berkeley: University of California Press, 1981); Peter Bowler, *Evolution: The History of an Idea*, rev. ed. (Berkeley: University of California Press, 1989), 301; Gregory Claeys, "The 'Survival of the Fittest' and the Origins of Social Darwinism," *Journal of the History of Ideas* 61 (2000): 223–40. A few scholars, however, emphasize Darwin's abolitionist sentiments and sympathy for other races, e.g., Greta Jones, *Social Darwinism and English Thought: The Interaction between Biological and Social Theory* (Sussex: Harvester Press, 1980), 140; Paul Crook, *Darwinism, War, and History: The Debate over the Biology of War from the 'Origin of Species' to the First World War* (Cambridge: Cambridge University Press, 1994), 25–28.

13. Darwin, *Descent of Man*, 1:240.

14. Ernst Haeckel, *Natürliche Schöpfungsgeschichte* (Berlin: Reimer, 1868), frontispiece, 555.

hierarchy he sketched did imply that Jews were biologically and intellectually inferior to Germans. He placed both Germans and Jews in the "Mediterranean species," which he considered higher than the other eleven "species." However, within the "Mediterranean species," he thought the Indo-Germanic race was superior to the "Semitic" race. He even thought they were different enough that they must have evolved from two different branches of ape-men.[15]

Daniel Gasman, in his zeal to argue that Haeckel was the progenitor of Nazism, overstates the extent of Haeckel's anti-Semitism, claiming that it decisively influenced Nazi ideology. However, the only evidence of Haeckel's anti-Semitism that Gasman adduces is one brief passage from his best-selling *Welträtsel* (1879, The Riddle of the Universe), in which he defended the idea that Jesus was partly "Aryan," and an interview with the Jewish journalist Hermann Bahr.[16] To be sure, both sources do show that Haeckel was anti-Semitic, but his rare remarks about Jews probably exerted little influence on his contemporaries. Rather, the lines of influence seem to run in the other direction. Haeckel admitted to Bahr that many of his most intelligent students were anti-Semitic (one of his students, Willibald Hentschel, was a rabid anti-Semite), so he could not overlook anti-Semitism.

In his interview with Bahr, which focused entirely on anti-Semitism, Haeckel expressed the view that anti-Semitism had some validity, but he overtly rejected the cultural, social, and religious forms of anti-Semitism. Rather, he stated, anti-Semitism "is a racial question." However, he then stated that if Jews assimilated to German culture, they should be tolerated, but not otherwise. Thereby, he elevated culture above race in determining how to treat Jews. He made a similar point in his discussion of the Eastern European Jews, against whom he favored immigration restrictions, stating: "Here a false humanity can only be harmful, and I think that we must energetically defend ourselves against the Russian Jews, not because they are Jews, but because they are irreconcilable with our culture [*Gesittung*]—like they protect themselves against the Chinese in California; the ideal love for 'everything that wears a human face' is unfortunately not practical to implement."[17] This statement seems to imply that the problem is not merely biological or racial, but rather cultural. Haeckel's position seems

15. Ernst Haeckel, *Natürliche Schöpfungsgeschichte*, 2nd ed. (Berlin: Reimer, 1870), 616–17.

16. Daniel Gasman, *The Scientific Origins of National Socialism: Social Darwinism in Ernst Haeckel and the German Monist League* (London: MacDonald, 1971), 157; Gasman, *Haeckel's Monism and the Birth of Fascist Ideology* (New York: Lang, 1998), 23–26.

17. Ernst Haeckel, interview in Hermann Bahr, *Der Anti-Semitismus: Ein internationales Interview* (Berlin: Fischer, 1894; reprint, Königstein: Jüdischer Verlag, 1979), 44–45.

rather ambiguous, but is intelligible if one understands that Haeckel (and many of his contemporaries) believed culture flowed directly from a particular race's biological traits. In his view, Jews who adopted German culture would be manifesting their biological superiority to Jews who did not assimilate.

Another leading Darwinist to emphasize the sharp racial distinction between Germans and Jews was the physician Ludwig Büchner, a famous scientific materialist who did more than anyone except Haeckel to popularize Darwinian theory in late nineteenth-century Germany. Büchner, like Darwin and Haeckel, believed that a wide variety of character traits—such as loyalty, diligence, thrift, laziness, greed, and deceit—were hereditary. In *Die Macht der Vererbung* (1882, The Power of Heredity), he argued that the moral character of nations and races is primarily hereditary, not cultural. Education and training cannot make a significant change in such character, and Büchner claimed that the constancy of the Jewish race proves the slow nature of such change. When Büchner specifically discussed the hereditary character of the Jews, he asserted that they were disposed to be merchants. Although he never specifically criticized Jews, one statement of his might hint at the inferiority of the Jews: "One can no more transform a savage into a civilized person, than one can make a Jew into a true Christian or a Semite into a Caucasian."[18] Büchner may or may not have been anti-Semitic, but he clearly treated the Jews as a race with mental and moral characteristics different from Germans.

Social Darwinist Racism among Sociologists and Anthropologists

Many late nineteenth-century German sociologists, anthropologists, and ethnologists began using Darwinian principles to explain human society and history, and race relations was a central feature of their social theories. One of the earliest German writers to apply Darwinism systematically to human social development was Friedrich von Hellwald in his magisterial *Culturgeschichte in ihrer natürlichen Entwicklung bis zur Gegenwart* (1875, The History of Culture in Its Natural Evolution), which impressed Haeckel. (Büchner contributed chapters to a later expanded edition.) Hellwald's vision of human history as an ineluctable Darwinian struggle between unequal races presaged Hitler's social Darwinist ideology. Hellwald argued

18. Ludwig Büchner, *Die Macht der Vererbung und ihr Einfluss auf den moralischen und geistigen Fortschritt der Menschheit* (Leipzig: Günthers Verlag, 1882), 49–50, on 59; see also Büchner's chapters in Friedrich von Hellwald, *Kulturgeschichte in ihrer natürlichen Entwickelung bis zur Gegenwart*, 4th ed. (Leipzig: Friesenhahn, 1896), esp. 1:101–2.

vigorously that the human struggle for existence is pitiless, stating, "The right of the stronger is a law of nature." Also, after stressing the inequality of races and warning that mixing "higher" and "lower" races would result in mediocre offspring, he stated: "Nature is and remains the greatest aristocrat, which mercilessly avenges every transgression against the purity of blood. Members of the same kind may mate only with those of the same kind, in human peoples (*Völker*) as in the animal and plant kingdoms."[19]

When Hellwald described the "lower" races in his writings, he most often mentioned blacks or American Indians. However, he also revealed anti-Semitic racism in *The History of Culture*. Like many other Darwinists, he believed that all races possess hereditary mental and moral characteristics, and in Hellwald's opinion, the character of "Semites" was deficient, at least when compared to the superior "Aryans." He charged that "Semites" do not possess a statesmanlike character and thus are unable to excel politically. He also denied that they have any creativity in science, technology, or the fine arts, except music. "In short," he concluded, "the Semite lacks most of what distinguishes the Indo-Germanic person."[20] Hellwald's amoral philosophy of racial competition shocked many of his contemporaries, but as Darwinism became more accepted, and as many European intellectuals gradually abandoned Judeo-Christian morality, Hellwald's brand of racism became more prominent until it was rather commonplace by the first decade of the twentieth century.

One social Darwinist writer who took Hellwald's brand of amoral racial struggle seriously was Alexander Tille, who wrote two books in the 1890s on the implications of social Darwinism for ethics and morality. Like Hellwald, Tille argued that Darwinism "recognizes no innate human rights" other than the right of the stronger. Rather the ethic that he derived from Darwinism focused on promoting human evolution.[21] One way to further human evolution was to introduce eugenic measures, which he zealously advocated. However, he also thought that racial struggle played a positive role in advancing the human species, stating, "Everywhere in nature the higher triumphs over the lower, and thus it is only the right of the stronger race to destroy the lower. When the latter do not have the ability to resist, they also have no right to exist."[22] Tille invoked Haeckel as his authority on

19. Friedrich von Hellwald, *Culturgeschichte in ihrer natürlichen Entwicklung bis zur Gegenwart* (Augsburg: Lampart, 1875), quotes on 27, 64–65.

20. Ibid, 134–35.

21. Alexander Tille, *Von Darwin bis Nietzsche: Ein Buch Entwicklungsethik* (Leipzig: Naumann, 1895), 23, quote on 204; Tille, *Volksdienst* (Berlin: Wiener'sche Verlagsbuchhandlung, 1893), 36.

22. Tille, *Volksdienst*, 26–27.

racial inequality and, like Haeckel, he usually stressed the supremacy of the "Mediterranean race" over the black Africans, American Indians, or Oriental races when he discussed the racial struggle. He did not mention Jews all that often in his writings, but when he did, he made clear that he considered them a branch of the "Mediterranean race" inferior to the Indo-Germanic branch.[23]

A prominent Jewish sociologist at the University of Graz, Ludwig Gumplowicz, contributed greatly to social Darwinist thought in the late nineteenth century with his book, *Der Rassenkampf* (1883, Racial Struggle). Gumplowicz thought history was dominated by the Darwinian struggle for existence between various human races. History, he asserted, is "the eternal lust for exploitation and dominance of the stronger and superior. The *racial struggle for domination* in all of its forms, in the open and violent, as well as in the latent and peaceful, is thus the essential *driving principle*, the *moving force of history*."[24] Unlike most anthropologists and social Darwinists of his time, however, he did not define races biologically, and he viewed racial mixture as a beneficent force in human evolution. Rather, he considered races to be social constructions, similar to what we would call ethnic groups today, as is evident in his discussion of the nature of the racial struggle:

> Struggle and war have *their particular compelling nature*, their particular bloodthirsty law that always and everywhere presses itself all-powerfully upon the ones struggling, and makes every struggle between heterogeneous ethnic and social elements into a "racial struggle," whether the antagonism between these races is large or small.[25]

Gumplowicz pleaded in vain against the rising tide of biological racism, which infected many scientists and physicians who appropriated his idea of racial struggle in the 1890s and thereafter. Although Gumplowicz would probably have been horrified by the way his ideas were used by subsequent racial theorists, Emil Brix is almost certainly right to conclude that Gumplowicz's amoral conflict theory "created one of the presuppositions for the pseudoscientific justification for the persecution of the Jews."[26]

Gumplowicz's conflict theory impressed many social Darwinists, including the American sociologist Lester F. Ward, who corresponded with him. His

23. Alexander Tille, "Der Kampf um den Erdball," *Nord und Süd* 80 (1897): 71.

24. Ludwig Gumplowicz, *Der Rassenkampf: Sociologische Untersuchungen* (Innsbruck: Wagner'sche Univ.-Buchhandlung, 1883), 218.

25. Ibid., 194.

26. Emil Brix, introduction to *Ludwig Gumplowicz oder die Gesellschaft als Natur*, ed. Emil Brix (Vienna: Harmann Böhlaus, 1986), 24.

most prominent disciple in the German-speaking world was the retired Austrian military officer Gustav Ratzenhofer, who wrote extensively on sociology. Gumplowicz even admitted that Ratzenhofer was a deep thinker, even a genius, who "stands high above me."[27] Ratzenhofer, however, seemed to take a more biological approach to race than did Gumplowicz. In applying the racial struggle to modern society, he saw colonial conquest and even extermination of "inferior" races as a natural and inevitable part of history, which would result in evolutionary advance. This process of racial extermination will continue, Ratzenhofer thought, until "all low-standing human races are destroyed, until finally the highest-standing Aryans remain in competition with the Jews."[28] There is little doubt who Ratzenhofer thought would triumph in the racial struggle between "Aryans" and Jews. Ratzenhofer's son, in a preface to his father's book, *Soziologie* (1907), divulged that his father had taken a keen interest in Houston Stewart Chamberlain's anti-Semitic ideas, and indeed several passages in *Soziologie* are strongly anti-Semitic.[29]

Anti-Semitic Theorists

Several prominent anti-Semitic theorists in late nineteenth-century Germany grabbed readily at the idea of a racial struggle for existence as the primary motive force in history. Unlike Gumplowicz, however, they defined races biologically and focused attention on the struggle between Germans and Jews. One leading anti-Semitic writer, Wilhelm Marr, in a sensational pamphlet published in 1879, invoked social Darwinian principles to justify a struggle against the Jews. His title pessimistically proclaimed *Der Sieg des Judenthums über das Germanenthum* (1873, The Triumph of the Jews over the Germans). As the subtitle of his pamphlet indicated, he did not consider the "Jewish question" a religious, but a racial or biological problem. In the past religion served as an excuse, but the real conflict, Marr thought, was "the fight of peoples (*Völker*) and their instincts against the *actual Judaizing (Verjudung) of society*, as a *struggle for existence*."[30] His attempt to put a cool, scientific spin on anti-Semitism led him to an amoral position. Neither Jews

27. Ludwig Gumplowicz to Lester F. Ward, 7 August 1902, in *Letters of Ludwig Gumplowicz to Lester F. Ward*, ed. Bernhard J. Stern (Leipzig: Hirschfeld, 1933), 10–11.

28. Gustav Ratzenhofer, *Positive Ethik* (Leipzig: Brockhaus, 1901), 319–20.

29. Gustav Ratzenhofer, *Soziologie: Positive Lehre von den menschlichen Wechselbeziehungen* (Leipzig: Brockhaus, 1907), pp. xii, 68–70, 228.

30. Wilhelm Marr, *Der Sieg des Judenthums über das Germanenthum*, 12th ed. (Bern: Costenoble, 1879), 13.

nor Germans were morally responsible for the struggle between them, since it was a product of inescapable biological laws. Further, he advised his fellow Germans not to hate the Jews, just as they do not hate individual enemy soldiers in wars: "The struggle between peoples (*Völkerkampf*) must be fought without hatred against the individuals, who are compelled to attack, as well as to defend themselves."[31]

Although the struggle motif would gain ascendancy in anti-Semitic circles by the beginning of the twentieth century, some prominent anti-Semitic theorists in the late nineteenth century rejected the "struggle-for-existence morality," as Eugen Dühring called it. Dühring preferred Lamarckian evolutionary theory to Darwinism, and he depicted the struggle for existence as an activity pursued by the allegedly immoral Jews. He claimed that Jews were bent on destroying other peoples, and Jews would annihilate themselves if confined to a Jewish state.

Despite his overt rejection of Darwinism, however, Dühring still founded his anti-Semitism on evolutionary theory. Like Haeckel, he promoted the idea that human races are separate species, and he considered some human races more advanced on the evolutionary ladder than others. Specifically, he judged the Asiatic peoples who founded the earliest empires in world history biologically inferior to the later Greeks and Romans. He also propagated biological racism, arguing that the mental and moral traits of the Jews are biologically determined. Like many later anti-Semitic racists, he claimed that culture is merely a reflection of biological character, so that peoples who create higher culture are also higher on the evolutionary scale. Lamarckian processes could not help the Jews or other lower races either, Dühring maintained, because evolution was so gradual that no significant change could occur within the foreseeable future.[32]

Despite Dühring's adoption of Lamarckian theory and his rejection of the struggle for existence in human society, when he described the long-term historical effects of racial differences, his description is strikingly similar to Darwinian theory. He asserted,

> [O]nly the more recent peoples have the complete aptitude and correspondingly also the calling to subject the Asiatics and to keep them externally as well as internally within limits, whereby they can only maintain a lower order of existence. They may continue existing, just like animals, as long as no sufficient reasons exist to completely displace them in this or that area. To

31. Ibid., 46.

32. Eugen Dühring, *Die Judenfrage als Racen-, Sitten- und Culturfrage*, 2nd ed. (Karlsruhe: Reuther, 1881), 109–17; Dühring, *Der Ersatz der Religion durch Vollkommeneres und die Ausscheidung alles Judenthums durch den modernen Völkergeist* (Karlsruhe: Reuther, 1883), 102–22.

> the extent that the better peoples on this planet need new land, they will press forward with their better species, and just as certain animal species, whose residence is not compatible with human management, retreat or perish where human culture expands, so also the advance and reproduction of the better human types will also limit the worse human types and cause them to disappear.[33]

For present society, Dühring's emphasis was on limiting the Jews through discriminatory legislation and policies. However, in the long run, he did seem to think that evolutionary processes would ultimately result in the extermination of "inferior" races, including the Jews.[34]

Obviously, neither Dühring, Marr, nor any other anti-Semite derived their anti-Jewish ideology from evolutionary theory. Rather they took anti-Semitic prejudices of long standing in German society and integrated Darwinian concepts into their racial ideologies. As many scholars have noted, anti-Semitism changed in the process; but this change was not, however, solely a result of Darwinism. Secularization in general required intellectuals either to abandon their anti-Semitic prejudices, as some liberals did, or to find a secular basis for their prejudices. Biological racism, which was often, but not always, closely intertwined with Darwinism in the late nineteenth century, provided a way to explain the pervasive prejudices without resort to religion. By viewing culture, including religion, as a product of biological, hereditary characteristics, some anti-Semites sought to explain the inferiority of the Jewish religion. Since biological racists, including many leading Darwinists, considered most moral characteristics hereditary, the common anti-Semitic caricatures of Jews as greedy, deceitful, sexually promiscuous, etc., could be explained as hereditary traits. Although some prominent racial anti-Semitic writers, such as Julius Langbehn, rejected Darwinism, most racial anti-Semites saw Darwinism as an integral part of their racial ideology.

Assimilation was no longer thought possible, much less desirable, by most of those who embraced this biological determinist view of race, as many prominent anti-Semites did. Some even warned that assimilation was dangerous. Converting to Christianity could not change Jews' biological disposition, so their deleterious, immoral traits would sneak into German society and culture. These secular anti-Semites thus discouraged conversion to Christianity, which they saw as part of a perfidious Jewish plot to infiltrate

33. Dühring, *Der Ersatz der Religion*, 122.

34. Reinhard Mocek, *Biologie und soziale Befreiung: Zur Geschichte des Biologismus und der Rassenhygiene in der Arbeiterbewegung* (Frankfurt: Lang, 2002), 181.

German society and corrupt German blood. Further, most racial anti-Semites considered miscegenation harmful, dragging down the superior race, so they discouraged intermarriage between Jews and Germans. Adolf Harpf, an anti-Semitic publicist who wrote some articles for Lanz von Liebenfels' *Ostara* journal, did not gain a significant following for his 1898 proposal to solve the "Jewish question" by intermarriage. Harpf, who was a proponent of Darwinian ethics, was a firm believer in biological racism and the Darwinian struggle for existence among races. Even though he was convinced that the "Aryans" are superior to the Jews in some ways, especially in moral character, his ethical impulses led him to reject persecution, discriminatory measures, or elimination as viable solutions to the racial struggle. Racial mixture, he thought, was the only moral way to avoid the destructive effects of racial competition.[35] Most racial anti-Semites were horrified by Harpf's proposal and thought it would lead to racial degeneration.

The Role of the Eugenics Movement

By the first decade of the twentieth century, Darwinized forms of anti-Semitic ideology became more widely disseminated. The burgeoning eugenics movement, which portrayed itself as Darwinism applied to society, strongly influenced the anti-Semitic movement.[36] Eugenicists promoted artificial selection as the way to overcome the allegedly deleterious effects of modern civilization, which was allowing the "unfit" to reproduce, thus leading to biological degeneration. Many eugenicists were convinced of the racial superiority of Europeans and saw eugenics as a way of winning the struggle for existence against other races. Alfred Ploetz, the key organizer of the German eugenics movement, which he dubbed the "race hygiene" movement, toned down his anti-Semitism and Nordic racism in public to win more followers, including Jews, from the medical profession. In his 1895 book promoting eugenics, he even claimed that Jews were mostly "Aryan" and were as high a race as the "Aryans," who were also a racial mixture.[37] However, in

35. Adolf Harpf, *Zur Lösung der brennendsten Rassenfrage der heutigen Europäischen Menschheit* (Vienna: Breitenstein, 1898), passim.

36. For a full explanation of the connections between Darwinism and eugenics, see Weikart, *From Darwin to Hitler*. Two works discussing anti-Semitism among German eugenicists are Robert Proctor, *Racial Hygiene: Medicine under the Nazis* (Cambridge: Harvard University Press, 1988); and Weindling, *Health, Race, and German Politics*.

37. Alfred Ploetz, *Die Tüchtigkeit unsrer Rasse und der Schutz der Schwachen: Ein Versuch über Rassenhygiene und ihr Verhältnis zu den humanen Idealen, besonders zum Socialismus* (Berlin: Fischer, 1895), 137–42.

private he warned his friend Gerhart Hauptmann to stop associating with Jews, because he thought they were exerting a deleterious influence on his playwriting.[38] He also secretly formed the Nordic Ring in Munich, a small organization devoted to spreading a Nordic racist form of eugenics. In a recruiting tract for the Nordic Ring he stated that "the Germans fight against other races a hard struggle for existence, which many view as hopeless." He, however, was not so pessimistic, but offered a solution. He explained that "for us the immediate foundation for the realization of our ideals can only be the Nordic race. . . . The object of our labor must be, in short, a Nordic-Germanic race hygiene."[39] Thus he encouraged his fellow Germans to promote eugenics as a way of improving the German race in order to win the racial struggle against other races.

Another branch of the eugenics movement was more overtly racist than Ploetz's. Ludwig Woltmann and the social-anthropological school that gathered around him and his doctrines became one of the leading forces promoting Nordic racism in early twentieth-century Germany. Woltmann and his colleagues viewed their work as a scientific enterprise, and Darwinism played a central, guiding role in all their writings. For Woltmann, race was the driving force behind all historical developments, and "the same process of natural selection in the struggle for existence dominates the origin, evolution, and destruction of the human races."[40] Woltmann's treatment of the racial struggle did not usually center on Jews, but on occasions he depicted Jews as inferior to the Nordic race.[41] Woltmann and his associates concentrated their efforts on "proving" through exhaustive research the superiority of the Nordic or Germanic race over all other races rather than attacking Jews. However, the implications of their Nordic racism were clearly inimical to Jews.

Woltmann gained a considerable following among racial thinkers, especially anti-Semitic Nordic racists, as Ludwig Schemann testified in his writings on racial theory. Schemann was the founder of the Gobineau Society and tirelessly promoted Gobineau's racial doctrines in Germany, but modified by social Darwinian thought and eugenics ideology. Though Schemann expressed a slight preference for Lamarckism over Darwinism, he was heavily

38. Alfred Ploetz to Gerhart Hauptmann, 28 November 1901, in Gerhart Hauptmann papers, Staatsbibliothek Preussischer Kulturbesitz, Berlin.

39. Alfred Ploetz, "Unser Weg," 1911, in Alfred Ploetz papers, privately held by Wilfried Ploetz, Herrsching am Ammersee.

40. Ludwig Woltmann, *Politische Anthropologie: Eine Untersuchung über den Einfluss der Deszendenztheorie auf die Lehre von der politischen Entwicklung der Völker* (Jena: Diederichs, 1903), 266.

41. Ibid., 307–9.

influenced by leading social Darwinists and eugenicists. He not only honored Woltmann, Darwin, and Galton as some of the greatest figures of the preceding generation (when he wrote in the 1930s and '40s), but his list of the influences on his own racial thought reads like a who's who of social Darwinists and eugenicists. The most important influence on Schemann's thought, by his own admission, was Otto Ammon, an engineer and amateur anthropologist, whose anthropology centered on the Darwinian struggle for existence among humans. One of Ammon's books was even entitled *Die natürliche Auslese beim Menschen* (1893, Natural Selection among Humans). In addition to Woltmann and Ammon, he also esteemed the works of Theodor Fritsch, Eugen Fischer, Hans F. K. Günther, Willibald Hentschel, Ploetz, and Wilhelm Schallmayer.[42]

Not all eugenicists embraced anti-Semitism. Some leading figures in the eugenics movement, such as Alfred Grotjahn and Felix von Luschan, were philo-Semitic. I have found no evidence that Schallmayer, a leading figure in the eugenics movement, was anti-Semitic, and he certainly opposed Nordic racism. Further, quite a few Jewish physicians, feminists, and sexual reformers embraced eugenics. As John Efron has demonstrated, leading Jewish anthropologists even embraced scientific racism in various guises in the early twentieth century, though they tried to use it to combat anti-Semitism.[43] In examining the prevalence of biological racism among Zionists, J. Doron has even surprisingly suggested that there "can be no doubt that many German-speaking Zionists esteemed as 'authorities' race ideologists like Gobineau, Chamberlain, Schemann, Wilser, Woltmann, Driesmans, Fischer, and Günther."[44]

Anti-Semitic Movement in the Early Twentieth Century

Nonetheless, many leading anti-Semites agreed with Ploetz that eugenics was a tool to win the racial struggle for existence. Theodor Fritsch, one of the leading anti-Semitic publicists in early twentieth-century Germany,

42. Ludwig Schemann, *Lebensfahrten eines Deutschen* (Leipzig: Matthes, 1925), 75, 295–97; Ludwig Schemann, *Die Rasse in den Geisteswissenschaften: Studien zur Geschichte des Rassengedankens*, vol. 3, *Die Rassenfragen im Schrifttum der Neuzeit*, 2nd ed. (Munich: Lehmann, 1943), pp. xi, xvi, 50, 249, 251, 256, 434.

43. John Efron, *Defenders of the Race: Jewish Doctors and Race Science in Fin-de-Siècle Europe* (New Haven: Yale University Press, 1994).

44. J. Doron, "Rassenbewusstsein und naturwissenschaftliches Denken im deutschen Zionismus während der Wilhelminischen Aera," *Jahrbuch des Instituts für deutsche Geschichte* (Tel Aviv) 9 (1980): 421. See also Weindling (chapter 5) and Falk (chapter 6) in this volume.

embraced a Darwinian worldview similar to Ploetz's. Like Ploetz and so many other eugenicists, he believed that the highest criterion for morality was the extent to which an action contributed to the preservation of the species. Whatever promoted the health and vitality of the highest members of the species was thus moral. He explained: "Morality and ethics arise from the law of preservation of the species, of the race. Whatever ensures the future of the species, whatever is suitable to elevate the race (*Geschlecht*) to ever higher stages of physical and mental perfection, that is moral."[45] Fritsch expressed roughly the same idea in many of his writings, including a book owned by the Nazi Party in Munich, wherein Fritsch asserted, "*Preserving the health of our race* is one of our highest [moral] commands."[46]

Fritsch spread his vision of Darwinian ethics widely through his publications, including his journal *Hammer*, as well as through his organization of the same name. In his *Handbuch der Judenfrage* (1887, Handbook of the Jewish Question), which by 1910 was in its twenty-seventh edition and thus was one of the most widely read anti-Semitic writings of the early twentieth century, he espoused scientific racism and often referred to the racial struggle against the Jews. He also integrated eugenic ideology into his work, vigorously promoting Willibald Hentschel's book, *Varuna*, which was published by Fritsch's Hammer Press in 1902. Fritsch admitted in the 1910 edition of his *Handbuch* that anti-Semitism needed to be founded on a complete worldview, and instead of developing that foundation himself, he pointed to *Varuna* as a work that elevated anti-Semitism to a science.[47] Fritsch also informed those who wrote for his journal, as well as his readers, "In broad outline the standpoint of the Hammer is characterized in Hentschel's work, *Varuna*, which counts as the programmatic statement of Hammer."[48]

Hentschel, who studied biology under Haeckel, opened *Varuna* by explaining that he was setting forth a scientific worldview underpinned by Darwinism. He expounded on Darwinian theory as the basis for his views on eugenics and race, stating: "The competitive struggle [between organisms]

45. Theodor Fritsch, "Die rechte Ehe: Ein Wort zum Züchtungs-Gedanken und Mittgart-Problem," n.d., in Reichshammerbund Flugblätter und Flugschriften, ZSg. 1-263/3, Bundesarchiv Koblenz.

46. *Vom neuen Glauben: Bekenntnis der Deutschen Erneuerungs-Gemeinde* (Leipzig, 1914), in Deutsche Erneuerungsgemeinde papers, ZSg. 1-263/6, Bundesarchiv Koblenz.

47. Theodor Fritsch, *Handbuch der Judenfrage*, 27th ed. (Hamburg: Hanseatische Druck- und Verlags-Anstalt, 1910), 6–7, 238.

48. Theodor Fritsch to Ludwig Schemann, 11 February 1908, in Ludwig Schemann papers, IV B 1/2, University of Freiburg Library Archives; Dieter Löwenberg, "Willibald Hentschel (1858–1947): Seine Pläne zur Menschenzüchtung, sein Biologismus und Anti-Semitismus," Ph.D. dissertation (University of Mainz, 1978), 46.

causes destruction of all those who are less vigorous or worse adapted—this is natural selection." While arguing that war is a positive selective force in history, he did not want to leave matters entirely in the hands of nature. Rather, he insisted that public policy should favor the "Aryan" race in its struggle against other races. He saw his task as "creating the foundations for a type of the Germanic race that will survive for the coming millennia—a new human creation."[49] Anti-Semitism is a constant theme in *Varuna*, as well as in Hentschel's other articles and books, most of which were published by Fritsch. Hentschel also promoted Woltmann's works on Nordic racism in anti-Semitic circles.[50]

The leaders of some Aryan supremacist cults in Vienna, which may have influenced Hitler's racial thought (though this is not certain), likewise integrated Darwinian race struggle and eugenics into their anti-Semitic ideology. In his journal *Ostara*, Lanz von Liebenfels often promoted eugenics and scientific racism. In an article, "Moses as Darwinist," for example, Lanz provided a mystical interpretation of the Bible (as he did in many articles), finding racial significance in scriptural passages that ostensibly have nothing to do with race. After revealing to his readers that the Trinity really represents three evolutionary stages of humanity, he stated: "Moses is thus actually a Darwinist, even a modernist, since evolution and selection are for him the driving force of all Being."[51] Earlier in the same article he asserted that the only reason the Jews had survived, despite being an inferior race, was because they followed the racial teachings of the Bible. However, he hoped that by appropriating the Bible, especially with his mystical racial interpretations, "Aryans" could outstrip all other races in the struggle for existence. He wanted "Aryans" to adopt the Bible "as the hard, racially proud, and racially conscious book, which proclaims death and extermination to the inferior and world domination to the superior."[52] To be sure, Lanz von Liebenfels did not overtly discuss Darwinism very often. Interestingly, he rarely discussed Jews, although both Darwinism and Jews featured prominently in this particular article. However, eugenics ideology and scientific racism pervaded his writings.

Despite his criticisms of Darwinian theory, Houston Stewart Chamberlain, one of the most famous anti-Semitic racial theorists of the

49. Willibald Hentschel, *Varuna: Das Gesetz des aufsteigenden und sinkenden Lebens in der Geschichte* (Leipzig: Fritsch, 1907), 14–15, 142.

50. Willibald Hentschel, *Vom aufsteigenden Leben: Ziele der Rassenhygiene*, 3rd ed. (Leipzig: Matthes, 1922), 5–6, 141.

51. Jörg Lanz von Liebenfels, "Moses als Darwinist, eine Einführung in die anthropologische Religion," *Ostara*, 2nd ed., no. 46 (1917): 16.

52. Ibid., 3.

early twentieth century, also synthesized some aspects of social Darwinism into his racial ideology. As a young man Chamberlain began training to be a biologist at the University of Geneva under the famous scientific materialist, Karl Vogt, who was one of the earliest proponents of Darwinism, but Chamberlain never completed his doctorate. As a student he zealously embraced Darwinism, by his own account, but later, when he adopted neo-Kantian idealism, he criticized Darwinism as too materialistic.[53] Nonetheless, he admitted that Darwinism did have some valid elements, especially the idea of selection and the racial struggle for existence. Chamberlain thus imported strong elements of Darwinian theory into his racial thought, even while remonstrating against Darwinism. The racial struggle for existence, which, according to Chamberlain, evolutionary theory accidentally discovered, is—even in its peaceful manifestations—a "struggle for life and death."[54] Because of these Darwinian elements in Chamberlain's worldview, it was very easy for other anti-Semites to adopt his main ideas, while rejecting his critique of Darwinism.

Early Twentieth-Century Anthropology

By the outbreak of the First World War, biological racism had become entrenched in anti-Semitic discourse and was also becoming mainstream among German anthropologists. In examining the anthropological literature in Germany relating to Jews, Georg Lilienthal has identified a significant increase in anti-Semitism around 1920, when the ideas of Eugen Fischer, Fritz Lenz, and Hans F. K. Günther gained prominence among German anthropologists.[55] However, as Fischer, Lenz, and Günther themselves admitted, most of their key ideas had already been advanced by earlier racial theorists and eugenicists, especially by Woltmann and his social-anthropological circle.

Fischer and his former student Lenz teamed up with Erwin Baur to publish a two-volume work on human heredity and eugenics in 1919. The largest part of the work, including the entire second volume on eugenics, was written by Lenz, who in 1923 was appointed professor of eugenics at the University of Munich. The text, published by Julius F. Lehmann, a leading publisher of medical texts, received many favorable reviews and was considered the standard text on human heredity in the 1920s and 1930s, not only in

53. Houston Stewart Chamberlain, *Lebenswege meines Denkens* (Munich: Bruckmann, 1919), 82–83.

54. Houston Stewart Chamberlain, *Die Grundlagen des neunzehnten Jahrhunderts*, 2 vols. (Munich: Bruckmann, 1899), 1:265–66, 278, 531; 2:717.

55. Georg Lilienthal, "Die jüdischen 'Rassenmerkmale': Zur Geschichte der Anthropologie der Juden," *Medizinhistorisches Journal* 28 (1993): 173–98, on 186.

Germany, but also in Britain and the United States, having been translated into English. Lenz explained to his friend Schemann that he had toned down his anti-Semitism in the first edition to help the book gain acceptance, and he crowed that his work had already been reviewed by five Jews; three were very favorable and two partly favorable toward his work. He promised Schemann, however, that subsequent editions would not be so favorable toward the Jews.[56]

The Baur-Fischer-Lenz text not only forcefully advocated eugenics and biological racism based on a Darwinian worldview, but it also promoted anti-Semitism, especially in the parts written by Lenz. Fischer's section on race from an anthropological perspective stressed racial inequality. When he turned his discussion to contemporary Europe, he acknowledged that both Germans and Jews are mixed races, but nonetheless he still thought they represented distinct racial types: "One can thus very properly speak of the racial characteristics and races of the Jews and Germans and can distinguish both sharply and clearly."[57] Fischer kept his anti-Semitic views veiled, cryptically remarking that the most important form of racial mixture presently occurring in Europe was the introduction of Asiatic racial elements by the Jews into European racial stock. Although Fischer never stated in his text that he disapproved of this miscegenation, most German anti-Semites would surely have viewed this racial mixture as harmful and would have taken Fischer's remarks as a warning.[58]

Indeed, Lenz was more explicit about his anti-Semitism in the Baur-Fischer-Lenz book than was Fischer, and in later editions Lenz forthrightly warned against mixed marriages between Germans and Jews. Already in the second edition he stressed the mental differences between races and claimed that the Jews had a special hereditary proclivity for commerce and transportation. Further, this biological predisposition gave them the ability to transmit culture, but not to create culture, as the Nordic race could. He also gave a scientific imprimatur to anti-Semitism by repeating standard stereotypes of the Jews as work-shy and manipulative.[59] In the fourth edition Lenz explicitly supported Nazi eugenics and racial policy, calling it applied biology. He specifically endorsed immigration restrictions aimed at Eastern European Jews. However, he also expressed regret for the "one-sided" anti-Semitism of

56. Fritz Lenz to Ludwig Schemann, 10 May 1922, in Schemann papers, IV B1/2, University of Freiburg Library Archives.

57. Erwin Baur, Eugen Fischer, and Fritz Lenz, *Grundriss der menschlichen Erblichkeitslehre und Rassenhygiene*, 2 vols., 2nd ed. (Munich: Lehmann, 1923), 1:148.

58. Ibid., 1:143.

59. Ibid., 1:407–25.

the Nazis, although it is not clear how sincere Lenz was in this regard.[60] Heiner Fangerau, who has analyzed the reception of the Baur-Fischer-Lenz text, is right when he states: "The criminal results of the National Socialist anti-Semitic terror find their origin in part in the BFL [Baur-Fischer-Lenz]."[61] Indeed, Hitler probably read the Baur-Fischer-Lenz book, possibly while composing *Mein Kampf* in prison. In any case, even before the Nazis came to power Lenz bragged that he had helped shape Hitler's worldview, claiming that "many passages in it [Baur-Fischer-Lenz] are mirrored in Hitler's expressions."[62] In 1933 Lenz reprinted an article he wrote in 1917, explaining in the preface that his earlier article set forth all the essential elements of the Nazi worldview.[63]

Hitler's Debt to Darwinism

Lenz's publisher, Julius F. Lehmann, was a close friend of Hitler in the early days of the Nazi movement in Munich, and his publications promoting social Darwinist racism, eugenics, and anti-Semitism probably had a significant influence on Hitler's worldview. By the 1890s Lehmann was a leading Pan-German nationalist, and in 1907-8 he became excited about eugenics as a means to rejuvenate the German nation. Toward the end of World War I Lehmann began publishing a magazine, *Deutschlands Erneurung* (Germany's Renewal), which promoted many ideas that found their way into Hitler's ideology. Whether it reflected ideas that Hitler already embraced from other sources, or whether it decisively influenced his thought is unclear. Lehmann also had close personal contact with Hitler, and he regularly gave Hitler copies of the books he published on eugenics and racism, many of which are still part of Hitler's library collection at the United States Library of Congress.[64]

Several of the books Lehmann presented to Hitler were by the notorious race theorist Hans F. K. Günther, who was appointed to a newly created chair

60. Erwin Baur, Eugen Fischer, and Fritz Lenz, *Menschliche Erblichkeitslehre und Rassenhygiene*, vol. 2, *Menschliche Auslese und Rassenhygiene (Eugenik)*, 4th ed. (Munich: Lehmann, 1932), 416–17, 504–5.

61. Heiner Fangerau, *Etablierung eines rassenhygienischen Standardwerkes 1921–1941: Der Baur-Fischer-Lenz im Spiegel der zeitgenössischen Rezensionsliteratur* (Frankfurt: Lang, 2001), 250.

62. Quoted in Ernst Klee, *Deutsche Medizin im Dritten Reich: Karrieren vor und nach 1945* (Frankfurt: Fischer, 2001), 256; see also Hans F. K. Günther, *Mein Eindruck von Adolf Hitler* (Pähl: von Bebenburg, 1969), 93–94.

63. Fritz Lenz, *Die Rasse als Wertprinzip, Zur Erneuerung der Ethik* (Munich: Lehmann, 1933), 5–7.

64. Phillip Gassert and Daniel S. Mattern, eds., *The Hitler Library: A Bibliography* (Westport, CT: Greenwood Press, 2001).

in social anthropology at the University of Jena in 1930 by the Nazi leader Wilhelm Frick, who was minister of education in Thuringia. In a brief account of the history of racial thought in the beginning of his book, *Rassenkunde des Deutschen Volkes* (3rd ed., 1923, Racial Science of the German People), Günther honored Darwin for encouraging later developments in racial anthropology. However, he claimed that the first really scientific work promoting the racial view of history was *L'Aryen, son Role Social* (1899, The Aryan's Social Role) by the social Darwinist Georges Vacher de Lapouge. Lapouge later joined the social-anthropological circle around Woltmann, another race thinker whom Günther greatly esteemed. Günther also thought highly of Ploetz, Schallmayer, and other eugenicists for their writings on heredity.[65] In a long appendix to his book on German ethnology he discussed the racial characteristics of the Jews. He repeated many common stereotypes, claiming, for example, that Jews lacked creativity and originality, except perhaps in the field of music. He also maintained that the racial antagonism that Germans feel toward Jews is not a cultural artifact but rather is rooted in the German's blood. This idea, that racial animosity was a biological instinct helping to preserve the race in the human struggle for existence, was a common theme in biological racism in the early twentieth century. Finally, he suggested the strict separation of Germans and Jews as the only viable solution to the racial problem. He called intermarriage *Rassenschande* (racial disgrace), a favorite term of the Nazis later when enforcing the Nuremberg Laws.[66]

Hitler probably read Günther's works in the 1920s, perhaps even before composing *Mein Kampf*. In any case, the social Darwinist and eugenic ideals espoused by Hitler in his only published book, as well as in many of his speeches, reflect the influence, either directly or indirectly, of Günther, Lenz, Woltmann, Fritsch, Hentschel, Chamberlain, Lanz von Liebenfels, and other racial thinkers of his time. As many scholars have explained, Hitler's worldview revolved around race. He viewed history as a Darwinian racial struggle, with the victors expanding at the expense of the losers. He spurned any moral codes—especially ones benefiting the weak and sick—that would interfere with the ability of the "Aryan" race to triumph in this struggle. He stated,

> If reproduction as such is limited and the number of births decreased, then the natural struggle for existence, which allows only the strongest and healthiest to survive, will be replaced by the obvious desire to save at any cost even the weakest and sickest; thereby a progeny is produced that must become ever

65. Hans F. K. Günther, *Rassenkunde des Deutschen Volkes*, 3rd ed. (Munich: Lehmann, 1923), 21–24.
66. Ibid., 487–91, 502–3.

> more miserable, the longer this mocking of nature and its will persists. . . . A stronger race (*Geschlecht*) will supplant the weaker, since the drive for life in its final form will decimate every ridiculous fetter of the so-called humaneness of individuals, in order to make place for the humaneness of nature, which destroys the weak to make place for the strong.[67]

His evolutionary ethic swept aside any humane impulses and provided a way to justify any action, no matter how abominable, if it promoted the interests of the "best" humans.

For Hitler the "best" humans were, of course, the "Aryans." He agreed with Lenz that they were intellectually superior to all other races and thus were the only real creators of civilized culture. Perhaps even more importantly, he considered "Aryans" morally superior, especially in comparison to Jews. While Hitler thought that "Aryans" were biologically disposed toward altruism, self-sacrifice, and honor, he depicted Jews as the epitome of evil and immorality. He accused them of greed, deceit, sexual deviance, and many other forms of immorality, all of which he considered innate, hereditary moral traits. According to his twisted vision of race relations, the reason Jews posed such a threat was precisely because they used clever, but deceitful, manipulation to take advantage of noble, but sometimes naïve, Germans. Since these immoral tendencies were allegedly inherited biological traits, the elimination of Jews from German lands was the only way to elevate morally the human species.[68]

In an unpublished 1928 manuscript, Hitler likewise clearly articulated a Darwinian vision of human society:

> While nature allows only the few most healthy and resistant out of a large number of living organisms to survive in the struggle for life, people restrict the number of births and then try to keep alive what has been born, without consideration of its real value and its inner merit. Humaneness is therefore only the slave of weakness and thereby in truth the most cruel destroyer of human existence.[69]

Hitler then defended infanticide for the disabled as the logical consequence of this view. He thus appealed to Darwinian principles to justify the right of the stronger to kill the weaker in the struggle for existence. Like many fellow Nazis, he used social Darwinist arguments to justify genocide against the Jews.

67. Adolf Hitler, *Mein Kampf*, 2 vols. in 1 (Munich: NSDAP, 1943), 144–45.

68. Ibid., 316–27.

69. Adolf Hitler, *Hitlers zweites Buch: Ein Dokument aus dem Jahr 1928*, ed. Gerhard L. Weinberg (Stuttgart: Deutsche Verlags-Anstalt, 1961), 56–57.

Conclusion: Darwinism and Anti-Semitism

Hitler, like many other anti-Semites in the early twentieth century, synthesized traditional anti-Semitic stereotypes into an overarching social Darwinist framework. Although I have concentrated in this essay on the social Darwinist forms of anti-Semitic thought, George Mosse has convincingly demonstrated that nonscientific, mystical forms of anti-Semitism also heavily influenced Nazi ideology. Dietrich Eckart, for example, a leading anti-Semitic figure in Munich whom Hitler considered his mentor, was not a social Darwinist. However, Mosse wrongly exonerated scientific racism, arguing that only the mystical form of anti-Semitism was murderous.[70]

Indeed, social Darwinist racism may have made anti-Semitism more murderous by contributing several elements to anti-Semitic ideology. First, Darwinism entailed biological variation within species, and many, including Darwin himself, thought this meant racial inequality. Second, the Malthusian population principle and the struggle for existence suggested that human populations were expanding faster than the food supply, so masses of humans would necessarily die in each generation, with only the fittest surviving. Third, social Darwinism contributed to the rising fear of biological degeneration by the end of the nineteenth century. Fourth, Darwinism provided the foundation for eugenics, which was a key ingredient in the Nazi worldview. Finally, biological racism increased substantially after the advent of Darwinism. To be sure, Darwinism did not necessarily imply biological racism (just as it did not necessarily imply eugenics), but nonetheless Darwinism gave impetus to hereditarian thought, and most biological racists were also avid Darwinists.

70. Mosse, *Toward the Final Solution*, 81–82; for further analysis of Mosse's position, see Richard Weikart, "Progress through Racial Extermination."

5

The Evolution of Jewish Identity: Ignaz Zollschan between Jewish and Aryan Race Theories, 1910–45

Paul Weindling

In the opening two decades of the twentieth century, a number of Jewish physicians, biologists, and anthropologists confronted the rising tide of anti-Semitic theories, which exploited science to assert the inferiority of the Jewish race. For these writers such racial theories were not only dangerously prejudiced but also scientifically erroneous. At the same time, they diagnosed the contemporary urban lifestyle as destructive to Jewish identity and sought through the rekindling of racial identity the regeneration and maintenance of Jewish identity in the modern world. These two concerns impelled them to try to establish a Jewish racial science. A key figure among these critics of German racial theory was the physician Ignaz Zollschan (1877–1948). Drawing on an evolutionary, although principally Lamarckian, paradigm for the development of races, Zollschan first outlined his views on race in *Das Rassenproblem* (1910).[1] His aim was to unite scattered Jewish populations by recognizing their common racial characteristics.

In his forcefully argued treatise on the *Rassenproblem*, Zollschan defined characteristics of a Jewish race by critically analyzing the secondary literature and scientific concepts. Unlike Rudolf Pöch in Vienna, who sought to map the racial characteristics of the skull forms in the Austro-Hungarian Empire, he did not conduct research into physical anthropology. Nor did he join any eugenics organization. He also differed from those Jewish race theorists who sought to work alongside advocates of German race regeneration (such as the anatomist Heinrich Poll during the 1920s) or who advocated

My thanks to the editors, to Veronika Lipphardt (Humboldt University Berlin) for critical comments and assistance with sources, to Christian Fleck (Graz) for access to the Hertz Papers, and to Hugh and Fred Iltis for information on their father, Hugo Iltis.

1. Ignaz Zollschan, *Das Rassenproblem unter besonderer Berücksichtigung der theoretischen Grundlagen der jüdischen Rassenfrage* (Vienna: Braumüller, 1910).

social hygiene (like the Vienna anatomist Julius Tandler and the Berlin sexologist Felix Theilhaber). However, having been an unwavering advocate of Jewish race purity during the ethnically diverse Habsburg monarchy before the First World War, from the later 1920s Zollschan played a leading role in an international coalition against Nazi racism. By that time he had radically revised his scientific ideas of a Jewish race, adopted an antiracialist stance, and played a major role in founding an international network of anthropologists to combat the threat of Nazi racism. The major focus of this paper is this transition in Zollschan's views about race and his political activities in confronting the threat of Nazism.

Ignaz Zollschan

Born in Erlach in Lower Austria, then part of the multiethnic Habsburg Empire, Zollschan was initially trained as a physician, qualifying in medicine in 1904 from the University of Vienna.[2] While a student in Vienna, which was governed by the anti-Semitic party, Zollschan had witnessed the intensification of racial anti-Semitism. In reaction he, like many other Jewish students, increasingly turned to Zionism and advocated the immediate settlement of Jews in Palestine.[3] Having completed his studies in Vienna, he worked as a ship's surgeon, thus enabling him to witness firsthand racial variations around the world. He subsequently returned to Vienna to train in the new specialty of radiology, one of the medical specialties that, like dermatology, offered openings for Jews.

Two years after *Das Rassenproblem* was published, he moved in 1912 to the historic spa resort of Karlsbad (Karlovy Vary) in the Bohemian region—referred to as the Northern Sudeten region by German nationalists—of what would soon become Czechoslovakia.[4] This area was a hotbed not only of anti-Semitism but also of anti-Czech sentiments among the belligerent German majority.

After World War I, Zollschan's views on Zionism changed along with his views on both race and Jewish identity, as he became increasingly critical of nationalism and racial thinking.[5] His interest in Jewish medical organizations

2. University of Vienna Archives, Promotionsakten.

3. Steven Beller, *Vienna and the Jews 1867–1938: A Cultural History* (Cambridge: Cambridge University Press, 1989), 201–3.

4. Elizabeth Wiskemann, *Czechs and Germans*, 2nd ed. (London: Macmillan, 1967), 101, 163, 179–80.

5. S. Winniger, *Grosse jüdische National-Biographie*, 7 vols. (Cernauti: Tipografia ARTA, 1931–32), 6:369.

in Palestine prompted a visit in 1927–28, when he worked as a radiologist at the Rothschild Hospital in Jerusalem and for Hadassah, the American-sponsored medical organization. He delivered lectures at the Hebrew University in the (unfulfilled) hope of obtaining an academic niche there.[6] Upon returning to Karlsbad he continued his radiological work but, with the increasing influence of the Nazis, he recruited anthropologists from the U.S.A., France, and Britain for an international network to oppose the errors of Nazi racism. In 1939 Zollschan lost his left arm due to his X-ray research, forcing him to retire from medical work. He soon moved to London, where he continued to pursue strenuously his anti-Nazi activities.[7]

Historiographical Issues

Zollschan needs to be located in the context of race theory, which was developing scientifically as well as becoming politicized. In the 1870s the liberal pathologist Rudolf Virchow had shown that Jewish racial characteristics were subject to considerable variation: there was a high incidence of blond hair and blue eyes among German Jewish schoolchildren. Yet by the turn of the century, the idea of finding a common index for a Jewish race proved attractive not only to anti-Semites but also to promoters of a secular Jewish identity. Jewish anthropologists hoped that by defining common characteristics of a race they would be taking the first therapeutic stage towards regeneration.[8]

At the turn of the century, Jewish and German race hygienists saw themselves as pioneers in mutually reinforcing enterprises. The Racial Hygiene Society, founded in Berlin in 1905 (becoming the German Society for Racial Hygiene in 1910), had both Jewish and socialist members. Among these were the dermatologist Alfred Blaschko, who had published on Weismann's germ plasm theory and class conflict in the socialist periodical *Die neue Zeit.*[9] The Zionist Arthur Ruppin competed for the Krupp Prize of 1900 for an essay on

6. Central Zionist Archives, Jerusalem (hereafter CZA), Ignaz Zollschan Papers, A 122/13, curriculum vitae: A 122/11/2, Jewish Agency for Palestine to Zollschan, 3 July 1940; Walter Zwi Bacharach, "Ignaz Zollschans 'Rassentheorie,' " *Jahrbuch des Instituts für Deutsche Geschichte* 6 (1977): S. 179–97.

7. CZA, A 122/11/1, Zollschan to Chief Rabbi J. H. Hertz, 5 April 1944.

8. Andrew Zimmerman, "Anti-Semitism as Skill: Rudolf Virchow's Schulstatistik and the Racial Composition of Germany," *Central European History* 32 (1999): 409–29; Constantin Goschler, *Rudolf Virchow: Anthropologe-Politiker* (Cologne: Böhlau, 2002).

9. Alfred Blaschko, "Bemerkungen zur Weismann'schen Theorie," *Die neue Zeit* 13 (1894): 19–23; Blaschko, "Natürliche Auslese und Klassentheilung," *Die neue Zeit* 13 (1894–95): 615–24; Paul Weindling, *Health, Race, and German Politics between National Unification and Nazism* (Cambridge: Cambridge University Press, 1989), 102–3.

race and nature; his treatise on Darwinism and social science won second prize.[10] Felix Theilhaber analyzed the declining birth rate among German Jews, as well as assessing the inhabitants of Berlin. He likewise considered that modern culture was causing the degeneration and extinction of Jews as a race, but he looked to a new Jewish nation for biological regeneration.[11] George Mosse, who extended the dimensions of racism to include ideas of health and sexuality, located Jewish doctors and biologists within the paradigm of a highly elastic Germanic ideology.[12] They were, he argued, Germans who happened to be Jewish, rather than the builders of a new Jewish race science.

But once the Germanic tradition was laid bare, was there also a Jewish race science tradition? John Efron has answered this question by showing how "Jewish race science texts" created "a new, 'scientific' paradigm and agenda of Jewish self-definition and self-perception."[13] While the example of Zollschan provides "the clearest and most unambiguous expression of Jewish race science," it is necessary to locate his ideas within broader debates on human evolution.[14] Despite Efron's pathbreaking analysis, differences among eugenicists with a Jewish background remain curiously invisible. This chapter will focus on Zollschan's distinctive and evolving position. He argued that culture was itself an evolutionary force, and that cultural values were more important than the varying physical characteristics. Eastern European Jews faced poverty and persecution while they retained a strong Jewish racial identity. By contrast, Western European Jews could advance in modern society, but this resulted in the destruction of their racial characteristics. For Zollschan, it was necessary not only to take a stand against anti-Semitic racial defamation and the abusive use of scientific methods, but also to strengthen Jewish culture in order to ensure that Jewish identity would be sustained.

10. Arthur Ruppin, *Darwinismus und Sozialwissenschaft* (Jena: Fischer, 1903); A. Beir, "Arthur Ruppin: The Man and His Work," *Leo Baeck Institute Yearbook* 17 (1972): 117–42; Weindling, *Health, Race, and German Politics*, 116–17.

11. Felix Aron Theilhaber, *Der Untergang der Deutschen Juden: Eine volkswirtschaftliche Studie* (Munich: Reinhardt, 1911); Theilhaber, *Das sterile Berlin* (Berlin: Marquardt, 1913).

12. George Mosse, *The Crisis of German Ideology* (New York: Grosset and Dunlap, 1964); Mosse, *Toward the Final Solution: A History of European Racism* (London: Dent, 1978); Mosse, *Nationalism and Sexuality: Respectability and Abnormal Sexuality in Modern Europe* (New York: Fertig, 1985).

13. John Efron, *Defenders of the Race: Jewish Doctors and Race Science in Fin-de-Siècle Europe* (New Haven: Yale University Press, 1994), 4–5.

14. Ibid.; Mitchell Hart, "Racial Science, Social Science, and the Politics of Jewish Assimilation," *Isis* 90 (1999): 268–97; Hart, *Social Science and the Politics of Modern Jewish Identity* (Stanford: Stanford University Press, 2000); Sander Gilman, *The Jew's Body* (New York: Routledge, 1981).

Zollschan's *Das Rassenproblem*

In *Das Rassenproblem* Zollschan tackled the "race question" in general, as he considered that it was impossible to treat the "Jewish racial problem" in isolation. The first part of his monograph was anthropological, the second historical and physiological, and the final section cultural. Drawing on Lamarckian environmentalism, this last section evaluated ideas of Germanic superiority and inherent Jewish inferiority, not from the point of view of political polemic, but from an empirical understanding of evolution and of history. At issue was the role of the Jewish race in history and in culture.

The basis of anti-Semitism had shifted from religion to race through the influence of anti-Semites like Eugen Dühring, Ernest Renan, Richard Wagner, and Houston Stewart Chamberlain. Zollschan outlined his definition of the Jewish race in response to Chamberlain, who advocated Aryan superiority on a theological and biological basis as a new form of German Christianity.[15] Chamberlain's *Die Grundlagen des neunzehnten Jahrhunderts* (1899, Foundations of the Nineteenth Century) saw the racial struggle in religious terms, with Jesus as a Teutonic redeemer. Cosima Wagner and Kaiser Wilhelm II were enthusiasts for this gospel of Aryan racial superiority.[16] Chamberlain lived in Vienna between 1889 and 1909, dabbling in botany, before moving to Wagner's Bayreuth.[17] Despite their fundamental differences, Zollschan agreed with Chamberlain that race was the fundamental category.[18] Zollschan warned against using physical anthropology of hair and skin color as racial markers and cautioned that the caricatured feature of the bowed Jewish nose was rare. Evolution thus provided a corrective to anti-Semites, and Zollschan's Lamarckism meant that culture and psychological identity were essential features of the Jewish race.[19]

Anti-Semites attributed the superior characteristics of the British, Germans, and Romans to the Indo-Germanic racial heritage, while Semitic Jews were viewed as inherently inferior and corrupting. Racial characters were assumed to be immutable. Zollschan roundly condemned the idea that all cultural achievement was Aryan. He cited the view of the anthropologist Felix von Luschan that there was no evidence for an Indo-Germanic race

15. G. C. Field, *Evangelist of Race: The Germanic Vision of Houston Stewart Chamberlain* (New York: Columbia University Press, 1981), 188–89; CZA, A 122/4/1, Zollschan draft letter to Chamberlain, n.d.

16. Zollschan, *Rassenproblem*, 220.

17. Ibid., 152–53, 205.

18. CZA, A 122/4/1, Zollschan draft letter to Chamberlain, n.d.; Chamberlain card to Zollschan, 17 December 1909.

19. Zollschan, *Rassenproblem*, 62, 96, 99–102, 135, 155.

with a common language and culture.[20] Zollschan rejected language as a racial characteristic, thereby attacking the resurgent interest in the Aryan theory of the French aristocrat Arthur de Gobineau, who had opposed Darwinism. Zollschan pointed out that Gobineau's cultural and linguistic approach was fundamentally different from that of physical anthropology. At this time a new wave of political anthropology was emerging, associated with a socialist Darwinist, Ludwig Woltmann, and conservative anthropologists like Eugen Fischer in Freiburg. Zollschan's intervention was timely. He pointed out the absurdity of postulating that such diverse languages as Persian and Scandinavian shared common Indo-Germanic roots. He also considered that theories of racial and ethnic conflict were polluting international life by chauvinistically setting race against race.[21]

In classifying Jews Zollschan adopted the Darwinist Thomas Henry Huxley's view of a predominance of a leukoderme type in Europe—divided between the "fair whites" (Xanthochroi), predominating in Northern Europe, and "dark whites" (Melanochroi) in Western and Southern Europe, North Africa, and the Middle East. Yet while Zollschan categorized Jews as belonging to the Melanochroi, he denied that cultural types were associated with these physical divisions.[22] Zollschan's Lamarckism meant that culture and psychological identity were essential features of the Jewish race, and evolution could thus provide a corrective to anti-Semitic views based on immutable physical characteristics.[23]

In *Das Rassenproblem* Zollschan refuted many negative stereotypes of Jews and the accusation that Jews damaged their host countries. He pointed out that while some authors saw Jews as inferior, incapable of achievement in sciences like mathematics, others accused Jews of intellectual superiority. Zollschan also denied any link between Jews and capitalism or between Jews and financial corruption, as anti-Semites had claimed, by noting that Jews had historically been excluded from the capitalist economy, and had often fared worst in any economic downturn.[24]

Zollschan was convinced that "at the present time, the destruction of the Jewish race is under way."[25] He believed that the Jewish race was in the throes

20. Ibid., 5.

21. Ibid., 8, 32.

22. Ibid., 144–45; T. H. Huxley, "On the Methods and Results of Ethnology [1865]" and "The Aryan Question and Pre-historic Man [1890]," in *Man's Place in Nature* (London: Macmillan, 1911), 209–52, 271–328; Huxley, *Science and the Hebrew Tradition* (London: Macmillan, 1901).

23. Zollschan, *Rassenproblem*, 62, 96, 99–102, 135, 155.

24. Ibid., 446–47.

25. Ibid., 155.

of dissolution and raised the question of whether it could exist in the future. His book stressed the role of environmental factors in determining culture. Viewing this process from a Lamarckian evolutionary standpoint, he argued that the Jewish race was degenerating because of environmental disadvantages. For example, only after the onset of assimilation did Jews suffer from syphilis and the resulting progressive paralysis. But even here Zollschan's position was more complex, since he maintained that on the whole the health of Jewish migrants had improved.[26]

Zollschan shared a common cause with the Zionist activist and psychiatrist Max Nordau, who responded to his conviction that Jews were degenerating by advocating that Jews build up their physique and muscular strength.[27] Nordau had observed how the educated Jews of the West would leave their religion, while those in the East would perish because of the miserable conditions of their existence. Likewise, Zollschan registered with dismay the large number of German Jews who converted, noting that numbers were lower in Russia. Ironically, the ghetto—despite its socioeconomic privations—was the best means of preserving the Jewish race.[28] Leaving the ghetto meant moving away from religious orthodoxy and a weakening of religiosity. Noting that Reform Judaism was a steppingstone to conversion and marrying out, he cited the high proportion of Berlin Jews who had abandoned their religion since the days of Moses Mendelssohn. The great Jewish intellectuals had drawn their inspiration from alien cultures, and they were the unconscious destroyers of their race.[29]

Zollschan espoused Zionism in the belief that by recognizing their racial identity Jews could be redeemed from the degenerative contingencies of the Diaspora. His conception of the Jewish race encompassed both Ashkenazi and Sephardi Jews, who shared a common racial type.[30] In supporting Theodor Herzl's Zionist vision of a new homeland, Zollschan believed that the racial individuality and culture of the Jews would be safeguarded, leading to the restoration of their self-respect and ethnic homogeneity.[31]

Zollschan argued that resolving issues in the theory of heredity was decisive for the racial problem, and focused on biological debates on instinct. Aware of the raging debate between Darwinians and Lamarckians, which had

26. Ibid., 268–70.

27. Peter Baldwin, "Liberalism, Nationalism, and Degeneration: The Case of Max Nordau," *Central European History* 13 (1980): 99–120.

28. Zollschan, *Rassenproblem*, 469.

29. Ibid., 486–89.

30. Ibid., 488–89.

31. Ibid., 421, 425, and 427.

convulsed fin de siècle biology, Zollschan favored environmentalist and associationist theories over the immutable hereditary particles of the Darwinist August Weismann.[32] He was convinced that Weismann's theory of a submicroscopic germ plasm did not explain adaptation and variation, and he praised Eduard von Hartmann's philosophy of the unconscious, which emphasized Lamarckian acquisition and provided a philosophy founded on biological instincts.[33] Zollschan saw instinct as providing a predisposition to reflective behavior, and it was therefore a characteristic of species and races and played a vital role in the struggle for survival.

For his views on the biological basis of inheritance, Zollschan was indebted to the biologist Richard Semon, a disciple of the Darwinist—but strongly Lamarckian—Ernst Haeckel. For Semon, repeated stimuli caused protoplasmic changes in nervous tissue; such traces he called engrams. The mneme was a composite of hereditary engrams and was the essential unit of psychological and biological characteristics over the generations. The mneme was racially specific, while any culture arose as an interaction between the inherited intellectual capacity and circumstances.[34] Zollschan appealed to the engram theory in responding to Otto Weininger's claim that the Jew is degenerate and effeminate. Instead, he explained genius by arguing that racial purity allows the instinctive basis of engrams a greater opportunity for expression.[35] Pure races were therefore intellectually more vital than racial mixtures; moreover, races needed to recover their primal vigor in order to sustain their creative capacities.[36]

Zollschan was also attracted by the Lamarckian-inspired experiments of the Jewish zoologists Paul Kammerer and Hans Przibram, in which gametes had been exposed to radium, thus providing important evidence for an environmental effect on the hereditary substance. Such Lamarckism allowed the psychology of the race to be modified, so that the Jew was not an immutable, fixed stereotype.[37] Semon offered technical corrections, but considered Zollschan's book important and timely. Zollschan was congratulated by other Jewish academics, such as the Viennese medical historian Max

32. Ibid., 223–25 and 232. On Weismann see Paul Weindling, *Darwinism and Social Darwinism in Imperial Germany: The Contribution of the Cell Biologist Oscar Hertwig (1849–1922)* (Stuttgart: Fischer, 1991).

33. Zollschan, *Rassenproblem*, 238.

34. CZA, A 122/3, Semon to Zollschan, 12 January 1911; Zollschan, *Rassenproblem*, 246–51, 294–97.

35. Ibid., 297–98.

36. Ibid., 420–21.

37. Weindling, *Darwinism and Social Darwinism.*, esp. 252–53; Arthur Koestler, *The Case of the Midwife Toad* (London: Hutchinson, 1971).

Neuburger. The British anthropologist A. C. Haddon also endorsed Zollschan's views.[38] With its blend of biology and cultural and religious history, Zollschan's book earned him academic respect. His views were further disseminated by the publication of American, English, and French editions of his writings.[39]

First World War and the Rise of Nazism

World War I was a turning point for Zollschan. In the midst of the political turbulence of closed borders, strikes, and pogroms during 1919, Zollschan announced his intention of purging nationalist elements from Zionism because nationalism was responsible for causing so much destruction during the recent war. He was perturbed by the increasing demands for political concessions to Jewish populations within the new nation-states—for example, Jews wanted official recognition of their status as a national minority in these states—and endorsement of those concessions by the Zionist movement. He believed that Zionists were playing into the hands of anti-Semites, who had long demanded special laws for Jews. In effect, this was throwing Jews back into the ghetto. Instead, he argued for internationalism on the basis of a Lamarckian theory of inheritance—that changes to ethnic identity could occur gradually, and this opened the possibility of rejuvenating the Jewish people in Palestine. He hoped that Zionism could be blended with internationalism and would uphold the ideals of international peace and cooperation.[40]

During the 1920s and early '30s, Zollschan was alert to the dangers of eugenics and increased his opposition to eugenics and to anti-Semitic racism. In 1925 he visited the Jewish anthropologist Franz Boas in New York to collaborate on X-ray investigations into the various races, having supported the use of X-rays to eradicate favus (a chronic skin infection) among East European Jewish children. Boas, who stressed culture over biology,

38. CZA, A 122/3, Semon to Zollschan, 22 October 1910, 12 January 1911; A 122/3, Max Neuburger to Zollschan, 26 October 1910; A 122/11/3, regarding Haddon's review of *Das Rassenproblem* in the *Morning Post.*

39. Ignaz Zollschan, *Jewish Questions: Three Lectures* (New York: Bloch, 1914); a revised fifth edition of *Das Rassenproblem* appeared in 1925. Translations of *Das Rassenproblem* appeared as *Le Rôle du Facteur Racial dans les Questions Fondamentales de la Morphologie Culturelle (Esquisse des Problèmes qu'aurait à se Poser une Enquête)* (Paris: Rousseau, 1934) and *The Significance of the Racial Factor as a Basis in Cultural Development* (London: Le Play House, 1934).

40. Ignaz Zollschan, *Revision der jüdischen Nationalismus,* 2nd ed. (Vienna: Braumüller, 1920), 127–59.

convened a committee at Columbia University that addressed human anatomical and psychological characteristics with the aim of refuting racist prejudice.[41] Zollschan subsequently used a memorandum drawn up by Boas in 1926 as a basis for intensified lobbying of leading intellectuals in Europe to refute anti-Semitic racism.[42] Later, in 1930, Zollschan published *The Significance of the Racial Factor as a Basis in Cultural Development* with Le Play House, a sociological research institution in London. Here he outlined such issues as the relations between Aryan and Nordic race theories and whether the Nordic claim that they possessed a superior culture was valid. The pamphlet made no mention of National Socialism and referred to Jews only in order to criticize the view that their cultural achievements arose from Nordic racial elements scattered among a number of other races, Jews included. Here Zollschan attacked the Nordic race theorist Hans F. K. Günther, as he had earlier attacked Chamberlain.

In 1933 the Prague Academy of Sciences organized a commission on race, and Zollschan also invited Thomas Masaryk, president of Czechoslovakia, to support an international coalition of notable intellectuals, anthropologists, and scientists opposed to the destructive implications of Nazi race theory.[43] The menace of Nazism likewise spurred Mendelian geneticists to make common cause with Lamarckians like Zollschan. His endeavors were paralleled by the Czechoslovak antiracist campaigner and socialist Hugo Iltis, the noted biographer of Gregor Mendel.[44] Iltis's critique of Nazi race myths as unscientific was based on Mendelism, in sharp contrast to Zollschan's commitment to Lamarckism. They also differed over tactics and religious outlook. Zollschan worked to establish national committees against unscientific racial

41. American Philosophical Society (hereafter APS), Boas Papers, Zollschan to Boas, 11 November 1925; Zollschan, "Die Favus-Ausrottungsaktion in Osteuropa," *Medizinische Klinik* 20 (1924): 100–102. Cf. Efron, *Defenders*, 166, 180; Doris Kaufmann, " 'Rasse und Kultur': Die amerikanische Kulturanthropologie um Franz Boas (1858–1942) in der ersten Hälfte des 20. Jahrhundert–ein Gegenentwurf zur Rassenforschung in Deutschland," in *Rassenforschung an Kaiser-Wilhelm-Instituten vor und nach 1933*, ed. Hans-Walter Schmuhl (Göttingen: Wallstein, 2003), 308–27.

42. APS, Boas Papers, Zollschan to Boas, 31 December 1933; Boas to Zollschan, 29 January 1934.

43. CZA, A 122/4/6, Memorandum to the Royal Society on the equality of the races of Europe. Cf. Ignaz Zollschan, *Zwei Denkschriften über die Notwendigkeit der Stellungnahme zum wissenschaftlichen Antisemitismus* (Karlsbad, 1933).

44. Hugo Iltis, *Gregor Mendel: Leben, Werk und Wirkung* (Berlin: Springer, 1924); Iltis, *Life of Gregor Mendel* (New York: Norton, 1932); Paul Weindling, "Central Europe Confronts German Racial Hygiene: Friedrich Hertz, Hugo Iltis, and Ignaz Zollschan as Critics of Racial Hygiene," in *Blood and Homeland: Eugenics in Central Europe 1900–1940*, ed. Marius Turda and Paul Weindling (Budapest: Central University Press, 2006), 299–321.

science—at the Academy of Sciences, Prague; in Vienna in 1937 under the dental anatomist Harry Sicher; in London under the Royal Anthropological Society and Royal Society; and through Franz Boas in the United States. His efforts gave rise to a Society for the Scientific Study of the Racial Question, founded in 1937. He hoped that an international panel would conduct a racial inquiry to evaluate various theories of race.[45]

Iltis adopted a secular outlook (he had a Jewish father, and while nominally Catholic was not religious) and was decidedly more populist in opposing racial theory than Zollschan, being active in education in Mendel's hometown of Brno (Brünn). In 1936 Iltis attacked the "poison gas" of Nazi race purity in his *Der Mythus von Blut und Rasse.* Calling for a popular mobilization against racism, he appealed particularly to Viennese public opinion, while also targeting academic anthropologists as a key Nazi group, not least because German anthropologists saw that Czechs could potentially be Germanized.[46] Iltis represented the Czechoslovak League against Anti-Semitism at the international congress to combat anti-Semitism held in Paris in 1937.[47] The more elitist Zollschan gained the support of the Czech political leaders, particularly the philosophically inclined Thomas Masaryk and Eduard Benes, who lobbied the supine League of Nations. Support also grew for the Ligue Internationale Contre l'Antisémitisme (LICA), which included Benes, Masaryk, and Einstein in its Committee of Honour.[48]

Both Iltis and Zollschan hoped that the avowedly antieugenic Catholic Church could become an ally against Nazi racism. Although there were German Catholic theologians in favor of eugenics, such as Hermann Muckermann and Josef Mayer, the condemnation of eugenics in the papal encyclical *Casti conubii* in 1930 gave rise to the hope, largely unrealized, that the church would take a firm stand against Nazi race ideology and policies. Zollschan approached Cardinal Innitzer, the archbishop of Vienna, and Iltis approvingly cited Cardinal Faulhaber's writings against the Aryan race theory. At a minimum they hoped to rein in the activities of anti-Semitic pastors, such as a notorious Pater Wilhem Schmidt in Vienna.[49] In pursuing this

45. CZA, A 122/4/1, I. Zollschan, "Die Bedeutung einer Rassen-Enquete" [typescript], Amsterdam 1938, A 122/4/10, concerning support from Gowland Hopkins and the Royal Anthropological Institute.

46. Hugo Iltis, *Der Mythus von Blut und Rasse* (Vienna: Harrand, 1936).

47. APS L. C. Dunn Papers, Iltis Correspondence folder 1: Iltis to Dunn, 9 February 1939; folder 4: Iltis to Mrs. Simon Guggenheim, 7 October 1951.

48. *Cahiers de la Licra* (Paris: Licra, nd); CZA, A 122 4/13, Einstein to Zollschan, 12 August 1936.

49. Hugo Iltis, *Volkstümliche Rassenkunde* (Jena: Urania, 1930); Iltis, *Mythus*; APS, Boas Papers, Zollschan to Boas, 28 April 1935.

goal Zollschan had two private audiences with Pope Pius XI, in 1934 and 1935.[50]

Zollschan's campaign against Nazi racialism led him to adopt the view that Jews were a culture rather than a race. From his Lamarckian perspective Zollschan inferred that biology and culture were fluid, not fixed, categories. As he embraced the view that Jewish identity is cultural rather than racial, he did not renounce his earlier ideas about the Jewish race; instead, he updated his views in response to Boas's theories and modern social science.[51] And as he focused increasingly on cultures, he found common ground with another Austrian (and nominal Catholic, albeit with a Jewish father), the social scientist Friedrich Otto Hertz (1878–1964), a pioneering critic of social Darwinism and eugenics who, like Zollschan, was more sympathetic to Lamarckism than to Mendelian notions of population genetics.

Hertz had formulated a seminal critique of scientific racism in 1902, and expanded this into two books, *Moderne Rassentheorien* and *Antisemitismus und Wissenschaft* in 1904, works that for the first time pointed to the social dangers of racial science.[52] Hertz's legal and sociological approach represented a very different perspective from Zollschan's scientific critique.[53] Hertz argued against associating fixed psychological types with different populations. Overall, he took the view that "race theories are little else but the ideological disguises of the dominators' and exploiters' interests."[54] There was no significant linkage between physical and mental characteristics. He found support from the eugenicist Wilhelm Schallmayer, but incurred the opposition of the Nordic racial hygienist and Mendelian Fritz Lenz.[55] Hertz

50. Efron, *Defenders,* 165. For background, see George Passelecq and Bernard Suchecky, *The Hidden Encyclical of Pope Pius XI* (New York: Harcourt Brace, 1997).
51. Zollschan, *Significance of the Racial Factor.*
52. Weindling, "Central Europe Confronts German Racial Hygiene."
53. Friedrich Hertz, *Moderne Rassentheorien: Kritische Essays* (Vienna: Stern, 1904). Hertz had studied law, and after a period in Austrian politics and administration was professor of political economy and sociology at the University of Halle in Germany from 1930 until 1933.
54. Friedrich Hertz, *Race and Civilisation* (London: Kegan Paul, 1928), 311.
55. Archiv Geschichte der Soziologie Österreichs, Graz, Nachlass Friedrich O. Hertz, Schallmayer to Hertz, 11 April 1915. Schallmayer sent reprints to Hertz on 13 March 1905 ("Selektionstheorie, Hygiene und Entartungsfrage"), and 1915 ("Unzeitgemässe Gedanken über Europas Zukunft"). Fritz Lenz, "Antwort an Hertz," *Archiv für Rassen- und Gesellschaftsbiologie* 12 (1917): 472–75; Lenz, "Friedrich Hertz. Rasse und Kultur. 3. Aufl. Leipzig 1925," *Archiv für Rassen- und Gesellschaftsbiologie* 18 (1926): 109–14; Friedrich Hertz, "Rasse und Kultur: Eine Erwiderung und Klarstellung," *Archiv für Rassen- und Gesellschaftsbiologie* 12 (1917): 468–72.

also found himself in opposition to the German anthropologist Eugen Fischer, who with the geneticist Erwin Baur and Lenz wrote the textbook of human heredity that the publisher J. F. Lehmann presented to Hitler in 1924. Hertz outlined his criticisms in *Race and Civilisation* (1928) and *Hans Günther als Rassenforscher* (1930).[56] These criticisms could readily be extended to Nazi racial ideology.

Zollschan's Role as International Organizer

During the 1930s Zollschan engaged in a number of attempts to organize a group of international experts to oppose scientific racism. In 1934 he approached Arthur Ruppin about establishing an institute for anthropological and sociological research at the Hebrew University. He sought finance from public associations—a Society for the Sociology and Anthropology of the Jews had just been founded in Vienna—as well as from such American foundations as the Carnegie trusts.[57] He was disappointed at the failure of these efforts. In the same year he persuaded the Royal Anthropological Institute and the Institute of Sociology to set up a committee to inquire into "the significance of race for cultural development."[58] The committee was a "broad church" encompassing not only Redcliffe Nathan Salaman and the anthropologist Charles Gabriel Seligman but also such non-Jewish academics as the social biologist Lancelot Hogben and the geneticists J. B. S. Haldane and R. A. Fisher. The committee's efforts were frustrated by squabbles among anthropologists over retention of the concept of race.[59] Zollschan found himself shunted into the antirace camp, while Salaman became disillusioned about the scheme for a bureau to inquire into "race values," and by 1936 he considered the spectrum of opinion too wide for obtaining any consensus.[60]

Nevertheless, Zollschan was gratified by the positive response of the Royal Anthropological Institute and Royal Society of London, and he continued his endeavors to establish an international coalition of scientific experts to refute

56. This was a translation of Friedrich Hertz, *Rasse und Kultur: Eine kritische Untersuchung der Rassentheorien* (Leipzig: Kröner, 1925).

57. APS, Boas Papers, Zollschan to Arthur Ruppin, 7 April 1934; I. Zollschan, "Bericht über die Vorarbeiten zur geplanten Rassenenquête," 1 October 1934.

58. CZA, A 122/3, Institute of Sociology, minutes, 4 May 1934.

59. Elazar Barkan, *The Retreat of Scientific Racism: Changing Concepts of Race in Britain and the United States between the World Wars* (Cambridge: Cambridge University Press, 1992), 285–89; Todd M. Endelman, "Anglo-Jewish Scientists and the Science of Race," *Jewish Social Studies* 11 (2004): 52–92.

60. CZA, A 122/11/4, Salaman to Zollschan, 23 July 1936.

the alleged scientific basis of Nazi race ideology.[61] In August 1937 he confronted head-on a team of Nazi race experts at the International Population Congress in Paris.[62] However, he was convinced that no individual was capable of examining the legitimacy of racial theory, since it was thought to constitute a monolithic "Wissenschaft in ihrer Totalität" (science in its totality). Instead, an international body of experts would be needed to expose the fallacies of Nazi race ideology. He worked strenuously to achieve an antiracist manifesto, which he hoped would be signed by Thomas Masaryk, Albert Einstein, Sigmund Freud, Aldous and Julian Huxley, the opponent of anti-Semitism James Parkes, and the novelist J. B. Priestley, among others.[63] Zollschan recruited academics in the hope of securing an international committee of nominated experts to evaluate the scientific basis of the theory of race.

The challenges Zollschan faced in his efforts to organize internationally are illustrated by his contact and correspondence with the skeptical historian of science and medicine, Charles Singer, who was greatly concerned by the rising tide of anti-Semitism. Singer, the son of a rabbi, was a key figure in discreetly rallying British academics to combat Nazism. In September 1936 he condemned Zollschan's scheme as impractical, undesirable, and dangerous. He argued that an international authority to refute Nazi racism was absurd because "scientific views are not established by international committees." Singer reflected that the experience of standardization of measurements proved that "[t]he whole history of science was against you" and that any international body was doomed to factionalism.[64] In responding to Zollschan's scheme for an international manifesto, Singer wrote that "as an English man of science," he opposed an international committee to define race as impractical, undesirable, and dangerous.[65] He was against the tyranny of any committee, and considered that the subject should be left to "free

61. CZA, A 122/4/6 [Zollschan], Memorandum to the Royal Society on the Equality of the Races in Europe.

62. Ursula Ferdinand, "Bevölkerungswissenschaft und Rassismus: Die internationalen Bevölkerungskongresse der International Union of the Scientific Investigation of Population Problems (IUSIPP) als paradigmatische Foren," in *Bevölkerungslehre und Bevölkerungspolitik im dritten Reich*, ed. Rainer Mackensen (Opladen: VS Verlag, 2004).

63. CZA, A 122/4/13, Zollschan, "Initiativcomité zur Veranstaltung einer Welt-enquête über die Rasssenfrage, An die Vertreter der Wissenschaft!," n.d. [typescript 1936]. The manifesto appears not to have been published.

64. CZA, A 122/11/5, Singer to Zollschan, 18 September 1936.

65. Cf. Tony Kushner, *The Holocaust and the Liberal Imagination: A Social-Cultural History* (Oxford: Oxford University Press, 1994).

research." A general solution to the race question could not apply to race in the United States and the racial position of Jews. However, he concurred with Zollschan's views on ethnology—recognizing culture and history as factors in framing Jewish identity.[66] Zollschan replied to Singer that he was primarily interested in refuting Nordic theories; an issue of vital importance to both Jews and non-Jews.[67] Faced by the worsening crisis after the Nazi annexation of Austria, Singer grudgingly conceded in April 1938 that "something may come of the idea."[68]

Zollschan was also in correspondence with the manic-depressive Julian Huxley, who suggested contacting leading authors, including his brother Aldous, H. G. Wells, and T. S. Eliot—blithely overlooking Eliot's expressions of disgust at Jewish peculiarities—for a public declaration on race.[69] Such prejudice among British intellectuals accounts for Singer's reticence. Singer continued to plead for discreet lobbying rather than public action. In July 1938 Hertz suggested to Walter Adams, who strongly supported refugess from Nazism, that a periodical should be launched to debate the issues of nationality and race. Instead Singer recommended a special discussion group at Chatham House.[70] By then British officialdom had liberalized the admission of refugees, much to the consternation of the Anglo-Jewish establishment, which advocated restrictive admissions policies.

In confronting Nazi anti-Semitism, Zollschan's endeavors were courageous, and it was Singer who capitulated. Singer may have been won over to grudging acceptance, but he was also cautious; after all, if science could undermine the concept of race in general, then Zionist theories of racial nationalism would also be undermined. In April 1938 Singer wrote, "[N]ow that the idea has started, the less both you and I appear in the matter the better. Nothing could be worse for its prospects than for it to appear to have behind it either a foreign or a Jewish motive power." He counseled Zollschan

66. CZA, A 122/11/4, Charles Singer to Zollschan, 18 September 1936; A 122/11/4, Julian Huxley, 19 September 1936; Wellcome Library, Singer Papers (PP/ CJS), Singer to Julian Huxley, 25 August 1942, in praise of Zollschan.

67. CZA, A 122/11/5, Zollschan to Singer, 21 September 1936.

68. CZA, A 122/11/5, Zollschan to Singer, 21 September 1936 and 16 March 1938; Singer to Zollschan, 18 September 1936, 17 March 1938, and 14 April 1938.

69. CZA, A 122/11/4, Julian Huxley to Zollschan, 19 and 29 September 1936; Aldous Huxley to Zollschan, 25 March 1936: Anthony Julius, *T. S. Eliot, Anti-Semitism, and Literary Form* (Cambridge: Cambridge University Press, 1995).

70. Bodleian Library, MSS SPSL, 351 f. 64 Hertz file, Walter Adams to Singer, 4 July 1938; f. 65, Singer to Adams, 25 August 1938. The Royal Institute of International Affairs is generally known as Chatham House.

regarding the anti-Zionist Board of Deputies: "In my opinion the Board of Deputies ought to take no action in this matter. The Board cannot do anything useful in this particular relationship, and any action that it took might be dangerous."[71] In contrast to the board's quiescence, Zollschan remained active in the public arena and continued to oppose racial theory when he moved to London in 1939.

Zollschan in London

In the British context, Zollschan's Czech nationality meant that he was accepted as a citizen of an allied nation, rather than having to suffer the discomforts of being an "enemy alien," as endured by Austrian Jewish refugees.[72] Zollschan, together with Hertz, who served as president of the Austrian Centre in wartime London, enjoyed a renaissance in popularity in the 1940s.[73] They were less concerned with the biological welfare issues preoccupying other interwar eugenicists. Being alert to the pernicious effects of racial science, they gained a new status at a time when liberal intellectuals realized the need for a common wartime front against biological racialism. Also, by this time, British biologists such as Julian Huxley, who supported Mendelian genetics and disapproved of Zollschan's reliance on Semon's engram theory, were prepared to condemn Nazi racism and to collaborate with those whom they disagreed with on scientific matters.

Despite the agreement to oppose Nazi racialism, there continued to be a lack of consensus on the question of Jewish identity. Hertz's sociology became linked to Morris Ginsberg's views and found expression in the papers on race and culture published in the *Sociological Review*.[74] Zollschan found a spectrum of opinions rather than unanimity on Jewish racial identity among British academics, both Jewish and gentile. For example in the antiracist tract, *We Europeans*, which Singer helped to produce, the Jewish problem was presented as cultural rather than racial: "The Jews do not

71. CZA, A 122/11/5, Singer to Zollschan, 14 April 1938. Cf. Geoffrey Alderman, *Modern British Jewry* (Oxford: Oxford University Press, 1992), 273, 281.

72. See *Bulletin of the Czechoslovak Medical Association in Great Britain* 2 (1942) for a collection of papers on the problem of race.

73. Hertz papers, correspondence relating to the Austrian Centre, Bodleian Library Oxford, Society for the Protection of Science and Learning (hereafter SPSL), 432 Friedrich Hertz file, for his London activism. For background see Marietta Bearman, Charmian Brinson, Richard Dove, Anthony Grenville, and Jennifer Taylor, *Wien-London, hin und retour: Das Austrian Centre in London 1939 bis 1947* (Vienna: Czernin, 2004), 16, 24–25.

74. Morris Ginsberg, *Reason and Unreason in Society: Essays in Sociology and Social Philosophy* (London: LSE and Longmans, 1947).

constitute a definite race, but a society forming a pseudo-national group with a strong religious basis and peculiar historical traditions. Biologically it is almost as illegitimate to speak of a 'Jewish race' as of an 'Aryan race.' "[75] Likewise, Huxley argued that "Jews of different areas are not genetically equivalent."[76]

The war increased the pressures for a rapprochement between Zollschan and key British figures like Julian Huxley and Salaman. For example, in 1940 Zollschan supported a meeting on "The Problem of the Jew" in order to intensify the battle against Nazi racism. It was the clear aim of the symposium to use "objective" anthropology, biology, and social science to define Jewish identity and to refute the stereotypes propagated by Nazi racial ideology.[77] The speakers were drawn from diverse positions. There was a substantial British Jewish contingent who strongly opposed racial ideology—Singer, Charles Seligman (the anthropologist), Morris Ginsberg (the sociologist), and Emanuel Miller (the pioneer of child guidance, and of the Jewish Health Organisation in 1930s London).[78] Yet some participants felt that if they rejected racial theory too dogmatically, their standing as experts might be compromised.

This episode came at the end of protracted efforts to modernize eugenics, discarding its imperialist racial hierarchies and adapting it to the modern welfare state, while disassociating it from the bogus and socially pernicious pseudoscience of Nazi racism. The pacesetters in this transition were the modernizing eugenicists Julian Huxley and the psychiatrist C. P. Blacker. The

75. Julian Huxley and A. C. Haddon, with A. M. Carr-Saunders, *We Europeans: A Survey of "Racial" Problems* (Harmondsworth: Penguin, 1935), 83. Cf. Barkan, *Retreat*, 296–97, 302–3.

76. CZA, A 122/4/13, "The Problem of Race: Julian Huxley's Call for an Inquiry," n.d., Royal Institution lecture; Julian S. Huxley, *Race in Europe* (Oxford: Oxford Pamphlets on World Affairs, 1939); Huxley, "The Concept of Race," in *The Uniqueness of Man* (London: Chatto, 1941), 116.

77. CZA, A 122/4/2, "The Problem of the Jew: A Scientific Symposium Arranged by the Jewish Scientific Institute (YWO). Section 1: The Race Problem, Subsections on the Anthropological Basis (with Fleure, Morrant, and Britzkus), the Historical Basis (with Huxley, Hertz and Singer), the Sociological Basis (with Seligman, Ginsberg, and Schwarzmann), the Biological Basis (with Haldane, Zollschan, Emanuel Miller); Section 2: The Jew in Literature; Section 3: The Jew in Economic Life."

78. Paul Weindling, "Jews in the Medical Profession in Britain and Germany: Problems of Comparison," in *Two Nations: British and German Jews in Comparative Perspective*, ed. Michael Brenner, Rainer Liedtke, and David Rechter (Schriftenreihe Wissenschaftlicher Abhandlungen des Leo Baeck Instituts 60) (Tübingen: Mohr Siebeck, 1999), 393–406. The small but innovative Jewish Health Organisation pioneered child guidance and social medicine, and exerted a positive influence more widely on welfare provision in London.

Anglo-Jewish scientists Salaman (a plant geneticist and author of the celebrated *History and Social Influence of the Potato* [1949]), his brother-in-law Seligman, and Singer were caught up in a transition from liberal imperialism—with Jews defined as a superior race—to liberal welfare policies, with Jews aligned with the superior elite, defined in terms of intellect and merit. Although Singer had taken part in an expedition to Abyssinia and Seligman had written *The Races of Africa* (1930), in the 1930s they opposed Nazi racialism.[79] Their intellectualist approach was also linked to their rejection of traditional Judaism, which they considered too formal and restrictive. For Singer universal science would come to replace religious superstition, and Nazi race science would likewise be undermined as possessing no intellectual credentials.[80]

In a tract entitled *Racialism against Civilisation* (1942), with a preface by Julian Huxley, Zollschan attacked "the Nazi theory of race" as a monolithic construct by an inhumane state dedicated to the destruction of civilized values. He argued that racism was not a problem that affected just the target group—the Jews—but was the common enemy to all religious, moral, and liberal political values. Zollschan continued to draw a distinction between "race theory" and "the scientific study of race," using the latter to expose the fallacies of the theory of an Aryan race.[81] Similarly, in 1943 Zollschan argued that the Nazi drive to force Jews back to the ghetto did not just represent a threat to the existence of Jews, but attacked the humanitarian basis of Western civilization. This represented a reversal of his earlier claim that the ghetto sustained Jewish racial identity.[82] However, the attack on "racial theory" weakened Zollschan's commitment to a Jewish race.

The efforts of Zollschan and other antiracist activists to combat the scientific premises of Nazi race theory were bearing fruit during the war. Although antiracist wartime propaganda has received surprisingly little attention from historians, it is apparent in scientific meetings, tracts, and lectures of the period. For example, one British Workers' Education Association pamphlet argued that Jews formed a cultural type and that the Nazi idea of a pure Jewish race was as flawed as the notion of an Aryan race.[83]

79. Barkan, *Retreat,* 30–34.

80. Geoffrey Cantor, "Charles Singer and the Early Years of the British Society for the History of Science," *British Journal for the History of Science* 30 (1997): 5–24, on 14.

81. Ignaz Zollschan, *Racialism against Civilisation with a Preface by Dr. Julian Huxley F.R.S.* (London: New Europe Publishing, 1942), 37.

82. Cf. Ignaz Zollschan, *Revaluation of Jewish Nationalism: A Sociological Study* (London: IGHL Zionist Fraternities of Austrian Universities, London Branch, 1943).

83. J. Irving, *Race, Nationalism, and Politics: An Outline for Study Circles* (London: WEA, 1940).

In correspondence, Huxley insisted that Zollschan should modernize his scientific assumptions by abandoning Semon's antiquated engram theory.[84] Although *Racialism against Civilisation* had been sponsored by Huxley, it had been rejected by the Jewish Defence Committee of the Board of Deputies of British Jews as likely to have only "a restricted reading public" and as being defective "from the point of view of style and English idiom."[85] Adolphe Brotman, the board's secretary, explained that Zollschan's work was "not exactly up our street."[86] The board, which had responded with excessive caution to the refugee problem, was reluctant to support a campaign against Nazi race ideology.[87] In April 1943 the indefatigable Zollschan outlined his strategy to Salaman: "As you know I have for nearly 25 years warned the Zionists of the danger of exaggeration. It must be feared that Diaspora nationalism[88] could never be accepted by leading British or American Jews." At that stage Zollschan's priority was—realistically—"saving Jews" and he wished to win over the leaders of Anglo-Jewry, such as Sir Robert Waley-Cohen, Leonard Stein, Lord Bearsted, Lord Melchett, Lord Samuel, James de Rothschild, and Sir Samuel Joseph.[89] The culmination of Zollschan's efforts came in May 1945 with a Conference of Allied Ministers of Education.[90] At this a statement[91] on the error of Nazi racial theories was prepared by Zollschan in collaboration with the leading geneticists C. D. Darlington, J. B. S. Haldane, and Julian Huxley. They condemned Nazi racial theories as a false religion.[92]

The Significance of Zollschan

John Efron and Stefan Kühl have portrayed Zollschan as failing in his attempt to set up an antiracist coalition. In his comparative analysis of

84. CZA, A 122/11/4, J. Huxley to Zollschan, 5 August 1941.

85. CZA, A 122/11/2, Jewish Defence Committee, Board of Deputies of British Jews to Zollschan, 18 December 1941 and 16 December 1942.

86. CZA, A 122/11/2, Brotman to Zollschan, 7 August 1942.

87. Alderman, *Modern British Jewry*, 281; V. D. Lipman, *A History of the Jews in Britain since 1858* (Leicester: Leicester University Press, 1990), 197.

88. "Diaspora nationalism" was the notion that Jews should claim recognition as a political minority in their countries of origin.

89. CZA, A 122/11/1, Zollschan to Salaman, 2 April 1943.

90. CZA, A 122/4/12, Report on Submission of Papers on Nazi Race Theory, Conference of the Allied Ministers of Education, Science Commission, *Year Book of the Inter-Allied Health Charter Movement* (1945).

91. Rice University, Julian Huxley Papers, J. G. Crowther, Conference of the Allied Ministers of Education to Julian Huxley, 24 April 1945.

92. CZA, A 122/11/6, Czechoslovak Republic, Ministry of the Interior, London to Zollschan, 27 April 1945; A 122/4/11, Conference of the Allied Ministers of Education Science Commission, Report on Submission of Papers on Nazi Race Theory, 12 June 1945.

changing concepts of race Barkan has recognized the value of Zollschan's lobbying during the 1920s and '30s and has emphasized the lack of a consensus over race until 1938—a difficulty that Zollschan sought to rectify. Yet Zollschan's achievements as a campaigner, critic, and idealist remain underrated. Prior to the Second World War his achievements included the creation of an international coalition against Nazi racism, which formed a basis for the UNESCO declarations on race after 1945.[93]

Zollschan's unrelenting campaign against racism can be compared with the struggle of Raphael Lemkin to gain recognition for the notion of genocide of culturally distinct peoples.[94] Such individuals were prescient and spoke from a mixture of firsthand observation of racial prejudice and a broad humanitarian vision. In his study of international eugenics Kühl observed that by the late 1930s the sympathizers with Nazi racism were reduced to a dwindling group, whereas the progressive American, British, and Scandinavian researchers argued for nonracist forms of eugenics. Kühl drew a distinction between the scientific advocates of eugenics and those anthropologists of Jewish descent, like Zollschan, Maximilian Beck, and Boas, who became increasingly antieugenic.[95] From this perspective, Zollschan's courageous response to Nazi racialism meant adopting the view that Jews were a culture rather than a race. In this he shared common ground with Hertz's sociology. As biology and culture were fluid categories within Zollschan's Lamarckianism, he could readily respond to the changing political environment and particularly to the Nazi threat to the existence of Jews.

The interactions between Zollschan and his British Jewish patrons reveal the strategies used to analyze and combat anti-Semitism. Singer and Zollschan shared a faith in positive science, but drew diametrically opposite conclusions. Singer did not wish to expose himself as a critic of Nordic racism, preferring behind-the-scenes activities. He scored some successes: the launch of the Academic Assistance Committee (later the Society for the Protection of Science and Learning), the book *We Europeans*, and a boycott by academics of the University of Heidelberg's anniversary celebrations.[96] But he had no wish to take a public stance as a Jew. Although Salaman was actively interested in the anthropology and demography of Jews, he concurred in putting the defense of humane values before that of Jewish

93. CZA, A 122, Zollschan to the editor, *The New Statesman and Nation*. Cf Special Collections, University of Sussex, New Statesman Archive, Editorial Correspondence 1944–1948, box 5.

94. Paul Weindling, *Nazi Medicine and the Nuremberg Trials: From Medical War Crimes to Informed Consent* (London: Palgrave, 2004).

95. Stefan Kühl, *Die Internationale der Rassisten* (Frankfurt: Campus, 1997), 153.

96. "Heidelberg, Spinoza, and Academic Freedom," *Nature* (22 February 1936), 303.

identity. These positions accord with the generally discreet and low-profile attitudes of British Jews.

In line with his Lamarckism, Zollschan adapted his ideas on race to the polarized sociopolitical context. He revised his views to accord with the expediencies of Jewish identity; shifting from a proeugenic to an antieugenic position. By 1946 Zollschan was arguing that Palestinian Arabs and Jews had common racial origins. By 1947 he opted for British naturalization in the hope of taking an active part in UNESCO's deliberations on race.[97]

The Zollschan who crossed swords with Houston Stewart Chamberlain in his *Das Rassenproblem* of 1910 was very different from the Zollschan who sought to create an international front against racial myths in the 1930s and '40s. He still argued that culture had a basis in biology, but this biology was itself infused by cultural values. Nazi racism diverted Zollschan from commitment to Zionism on a biological basis to the broader issue of race. His efforts came to focus on the building up of an international coalition of scientists against racial myths. After the First World War he abandoned nationalism. After the Second World War he abandoned race. His final posthumous publication no longer saw a "racial problem" but "racial insanity."[98]

97. CZA, A 122/4/2, Zollschan to Fleure (re naturalization), 21 March 1947; Zollschan to Sir Alfred Zimmern, UNESCO, 16 May 1946; *Statement on Race: An Annotated Elaboration and Exposition of the Four Statements on Race Issued by the United Nations Educational, Scientific, and Cultural Organization*, ed. Ashley Montagu (New York: Oxford University Press, 1972).

98. Ignaz Zollschan, *Der Rassenwahnsinn als Staatsphilosophie: Mit einem Vorwort von Julian Huxley* (Heidelberg: Lambert Schneider, 1949); CZA, A 122/4/7, Urexemplar booklet and pamphlet, typescript "The Pseudo-Science of Racism." The draft has three alternative titles: "Racism against Civilisation," "Revolution against Freedom," and "Racial Doctrine as a Political Philosophy."

6

Zionism, Race, and Eugenics

Raphael Falk

> Finally, allow me, gentlemen, to mention one other disease, not mentioned so far, which is yet very important. This is the ancient suffering that the composer Heinrich Heine called "the Jews' disease." In his famous verse at the occasion of the inauguration of the Jewish hospital in Hamburg he wrote
>
> *Ein Hospital für arme kranke Juden*
> *Für Menschenkinder, welche dreifach elend*
> *Behafet mit den bösen drei Gebresten*
> *Mit Armut, Körperschmertz und Judentume.*[1]
>
> And I ask: Is Jewishness a disease? Jewishness by itself is not a disease! The disease of the Jews is nothing but the reflex of the world's morality.—If, however, disease is suffering, then indeed there exists a Jews' disease, a very severe one.
>
> HERMANN ZONDEK, 1940[2]

In the late nineteenth and early twentieth centuries the principal aim of Zionism was to establish an autonomous national existence in the historic homeland for the Jewish people. It brought together Jews from diverse backgrounds. Those from Western Europe were often disappointed with the outcome of the emancipation movement and the failure of society at large to fully accept Jews. By contrast, Jews from Eastern Europe were usually seeking emancipation, while also hoping to maintain a distinctly Jewish way of life. As a sociopolitical movement Zionism was a latecomer

The author thanks Geoffrey Cantor and Marc Swetlitz for inviting him to contribute this paper and for their conscientious work in editing it.

1. A hospital for sick and needy Jews,
   For human beings, who are triply wretched,
   With three great maladies afflicted:
   With poverty, corporal pain, and Jewishness.
2. Hermann Zondek, "On Medicine and Jewish Physicians in Our Times" (in Hebrew), *Ha-Refu'ah* 18 (1940): 29–32.

among the European national movements that struggled for self-definition and independence.

In 1862 Moses Hess (1812–75), a socialist and precursor of Zionism,[3] published *Rom und Jerusalem: die letzte Nationalitätsfrage* (1862, Rome and Jerusalem: A Study in Jewish Nationalism), in which he claimed that "Jews are primarily a race that in spite of all the influences of climate adapted to all situations and maintained its integrity."

> The Jewish race [claimed Hess] is one of the primary races of mankind that has retained its integrity, in spite of the continual change of its climatic environment, and the Jewish type has conserved its purity through the centuries. The Jewish race, which was so pressed and almost destroyed by the many nations of antiquity, would have disappeared long ago in the sea of Indo-Germanic nations, had it not been endowed with the gift of retaining its peculiar type under all circumstances and reproducing it. . . . My own race has played such an important role in the world history and is destined for a still greater one in the future.[4]

Hess followed Giuseppe Mazzini's harmonist notion, combining national particularism with a universal vision: Mazzini said that by being a member of a nation, one is also a member of the human race, and that the only way of belonging to humanity is to belong to a specific nation.[5] However, Hess conceived the Jewish problem as a national problem, rather than one of equal rights and emancipation for a religious sect. On Hess's analysis, the failure of the Jewish Reform movement resulted from its tendency to view the Jewish problem solely in religious terms, thus ignoring the true historical essence of Jewishness with its material, biological roots—a *Volk* in the sense of Herder and Hegel.[6] He concluded that a national homeland in Palestine—rather than assimilation—was the proper resolution for the Jews.

Like other mid-nineteenth-century writers Hess used terms like "race," "nation," and "*Volk*" rather indiscriminately. However, in the second half of

3. Shlomo Avineri, *Moses Hess: Prophet of Communism and Zionism* (New York: New York University Press, 1985); Avineri, *Varieties of Zionist Thought* (in Hebrew) (Tel Aviv: Am Oved, 1980).

4. Moses Hess, *Rom und Jerusalem* (Leipzig: Eduard Wengler, 1862), quoted in Israel Eldad, "Jabotinsky Distorted," *Jerusalem Quarterly* 16 (1980): 23 (http://www.saveisrael.com/eldad/eldadjabo.htm, accessed 16 August 2005).

5. Shlomo Avineri, *The Making of Modern Zionism: The Intellectual Origins of the Jewish State* (London: Weidenfeld and Nicholson, 1981), 45.

6. There is no satisfactory translation of *Volk* (and its derivatives). In nineteenth-century Central European culture the concept incorporated diverse overlapping notions, such as nation, folk, race, language, common history, homeland, even tribe or clan.

the nineteenth century the life sciences became firmly established on materialist assumptions, especially after the publication of Darwin's *Origin of Species* in 1859. The increasingly successful evolutionary perspective significantly reshaped social and political thought. Social Darwinism, as expounded by Herbert Spencer, is perhaps the most explicit expression of the impact of materialist biological conceptions on social thought. Another outgrowth of the impact of materialist biology on social and political conceptions was eugenics. Of particular importance to the present discussion was the prominence given by social Darwinians and eugenicists to the inborn, hereditary element of *Volk* and race in contrast to elements that were culturally acquired.

In his book *Varieties of Zionist Thought* (1985) the political historian Shlomo Avineri suggested that the Zionist movement matured towards the end of the nineteenth century because of developments in the world at large, primarily stemming from the French Revolution. For the first time Jews in many countries attained the economic and social mobility that enabled them to integrate into the wider society. Yet, these changes were accompanied by a rise in romantic European nationalism, which, in the second half of the nineteenth century, was increasingly underpinned by biological-racial conceptions. Thus, to the extent that Jewish tradition had generated diverse modes of coping with specific predicaments, the opening of the liberal society to Jews created new tensions, which for the Jews turned emancipation into a dialectic process increasingly fueled by anti-Jewish sentiments.[7] Monika Richarz has appropriately described Jewish emancipation as "a shift from the status of a Protection Jew (*Schutzjude*) to that of a second-grade citizen."[8] Emancipation thereby led Jews to conceive of a Zionist nation-state as the only redemption of the Jewish people's plight, sociopolitically as well as sociobiologically.

Theodor Herzl's (1860–1904) disenchantment with the expectation that Jews would integrate into European society prompted him to engage in political action. Like other ethnic groups Jews possessed a common cultural and biological history and were therefore entitled to live in their own homeland. Herzl was convinced that by political, mainly diplomatic, means it would be possible to reestablish the Jewish people as a nation among nations. At a time of liberal and nationalistic fervor, Zionism offered an answer to the challenge of Jewish identity in terms of liberalism and nationalism. Zionism

7. Avineri, *Varieties of Zionist Thought*, 15–24.

8. Introduction, *Jewish Life in Germany: Memoirs from Three Centuries*, ed. Monika Richarz (Bloomington: Indiana University Press, 1991), 1.

therefore adopted the prevailing notion of a *Volk*, a nation-race that had been shaped by its blood and soil (*Blut und Boden*), and demanded that Jews be reinstated in their historic homeland.

Although eugenics and Zionism had completely different ideological roots, both were products of the materialistic beliefs that underpinned much social philosophy in the second half of the nineteenth century. Both articulated strong utopian programs.[9] While the former focused on the improvement (or prevention of the degeneration) of the human species, the latter addressed the future of the Jewish race. Both were based on the achievements of scientific rationality.

In the present paper I will show that many Zionist writers appealed to biological conceptions of race and nation and displayed an awareness of their responsibility not only to preserve this biologically circumscribed ethnic group but also to propagate and improve it. Although never a major issue in the complex history of Zionism, I will argue that it has been a persistent one. Before World War II the emphasis was primarily on overcoming those degenerate qualities that Jews were charged with having accumulated while living in the Diaspora. After the Holocaust and the gathering of exiles in the new State of Israel the focus changed to the search for common genetic denominators to Jewish communities dispersed throughout the world that would establish their ancient roots in the Land of Israel. Advances in genetic research endowed eugenics with a new significance.

Jews as a *Volk*

From its foundation, political Zionism aspired to establish a new Jewish type in its homeland. Implicitly, if not explicitly, Zionists accepted the anti-Semites' claim that the Jews were not only a race but also a degenerate one. The term "anti-Semitism" had been coined in 1879 by Wilhelm Marr (1819–1904) in his book *Der Sieg des Judenthums über das Germanenthum* (The Triumph of the Jews over the Germans). He alluded to the notion, prevalent after Darwin published his theory, that material evolution applied to human species. Accordingly, differences between peoples and their social characteristics had been shaped by natural selection. Marr's book was a deliberate attempt to show that discrimination against the Jews was based on "scientific" anthropological foundations: they were an alien *Volk* or nation.[10]

9. Herzl's Zionist utopia *Altneuland* (Leipzig: Hermann Seemann Nachfolger) was published in 1902.

10. Wilhelm Marr, *Der Sieg des Judenthums über das Germanenthum: Vom nicht confessionellen Standpunkt aus Betrachtet* (Bern: Rudolf Constable, 1879).

Within a year of its initial publication Marr's book passed through twelve editions!

In February 1882, when the twenty-two-year-old student Theodor Herzl learned about the enforced isolation of the Jews in medieval times, he noted that this had "prevented the physiological improvement of their race through crossbreeding with others." And, he continued, "[b]asically the Jews have a different physical and mental physiognomy . . . because they interbreed rarely with members of other nations. . . . Crossbreeding of the Occidental races with the so-called Oriental ones on the basis of a common state religion—this is the desirable great solution."[11]

Max Nordau (1849–1923), who became one of Herzl's earliest and most loyal followers, had by 1892 become an ardent promoter of the ideas of the decline of Western culture and its biological correlates. Nordau insisted that the "typical" Jewish mind and physique were degenerate. In 1896 he emphatically affirmed the decline of Jewry when responding to the question (posed by the editors of the *Allgemeine Israelitische Wochenschrift*): "Are Jewry and the Jews in the process of decline, and, if so, what means could arrest the process?" Both mentally and physically Jews showed symptoms of degeneration. Millennia of persecution and humiliating living conditions resulted in a depressive frame of mind. As a people, they have lost contact with the soil. "A people cannot, in the long run, remain healthy and strong if it does not again and again, at least temporarily, return to the rejuvenating soil. Without this Antaeus-treatment it inevitably falls prey to wasting disease."[12] Nordau recommended that Jews indulge in open-air physical activities, such as agriculture and sports. "Life in nature will rejuvenate their bodies, the secure possession of the soil will resurrect their self-esteem". He also addressed the origins of the characteristics manifested by Jews:

> A convincing and adequate scientific explanation is difficult to formulate, since the anthropology and ethnography of the Jewish stock is a nearly completely untouched field of research. Thus, we do not know if the Jews originally had a greater body length, so that they became stunted only as a consequence of their unfavorable living conditions, or whether they were a race of minor length to begin with.[13]

11. Sanford Ragins, *Jewish Responses to Anti-Semitism in Germany, 1870–1914: A Study in the History of Ideas* (Cincinnati: Hebrew Union College Press, 1980), 106.
12. Meir Ben-Horin, *Max Nordau: Philosopher of Human Solidarity* (London: London Jewish Society, 1956), 180. Antaeus, the mythic Greek giant wrestler, was invincible as long as he was in touch with his mother, the earth.
13. Max Nordau, "Was Bedeutet das Turnen für uns Juden?" [1902], reprinted in Nordau, *Zionistische Schriften* (Cologne: Jüdischer Verlag, 1909), 382–88, on 384.

Nordau was less interested in the historical roots of the Jews than in the influence of environmental conditions on their biological nature. His notions of race and nation were primarily taken from Herder and the German *Volksgeist,* which considered the nation as an ontological entity in its own right, having precedence over those of its individual members.

The best-known instance of a *völkisch* manifesto in the early history of Zionism was Martin Buber's 1911 celebration of "blood" as the paramount essence of Jewish identity. Buber argued that the Western Jew is rootless. A homeland, with its language and customs, was alien to his essential being, as they were not part of his "community of blood" (*Gemeinschaft seines Blutes*). Nevertheless, the Jews can become an "autonomous reality" because it means more than a mere continuity with the past: "it has planted something within us that does not leave us at any hour in our life . . . blood, the deepest, most potent stratum of our being." A religious philosopher, Buber conceived the Zionist idea in terms of both humanism *and* nationalism. Like a child "[s]tirred by the awesomeness of eternity, . . . when he envisions the line of fathers and of mothers that had led up to him," the Jew perceives "what confluence of blood has produced him. . . . He senses in this immortality of the generations a community of blood."[14] Buber did not generally think in biological terms, yet it is hard to avoid the conclusion that when he spoke of "blood," he was trying to establish the Jewish national identity in biological (though very much Lamarckian) terms: "[T]he innermost stratum of man's disposition, which yields his type, the basic structure of his personality, is that which I have called blood: that something which is implanted within us by the chain of fathers and mothers, by their nature and by their fate, by their deeds and by their sufferings."[15]

The most explicit contender among the Jewish—and certainly the Zionist—politicians who asserted the existence of a Jewish race was Ze'ev Vladimir Jabotinsky (1880–1940), a member of the Zionist Executive. After being imprisoned for leading the defense of the Arab anti-Jewish rioters in

14. Martin Buber, *Drei Reden über das Judentum* (Frankfurt am Main: Literarische Anstalt Rütten & Loening, 1923), 11–21. See Buber, "Judaism and the Jews" in *On Judaism by Martin Buber*, ed. Nahum N. Glatzer (New York: Schocken, 1967), 13–15.

15. Buber, *Reden*, 11. In the Hebrew edition of his original 1911 essay Buber added a postscript: "A couple of years later evil people misconstrued the concept of 'blood' that I made use of. Therefore, I herewith declare that in every place that I referred to blood, I under no circumstances meant the racial issue, which in my opinion is baseless, but rather to the continuity of procreation and births of a nation, which provides the backbone for its essence." Postscript in Hebrew to the Hebrew translation of "Das Judentum und die Juden" in *Te'udah ve-Ye'ud*, ed. M. H. Ben-Shamai et al. (Jerusalem: Zionist Library, 1984), 29.

Jerusalem in 1920, he left Palestine and in 1925 founded the Alliance of the Revisionist Zionists, which eventually resulted in the establishment of the New Zionist Organization in 1935. Jabotinsky was strongly affected by the nationalistic and futuristic ideas prevalent during his years as a student in Italy. He considered himself an uncompromising liberal democrat, and his ideological conceptions were highly deterministic. He asserted that only identifiable national entities can contribute creatively to the human spirit. To achieve this state of potentially creative nationality, a nation must secure proper conditions, such as a separate language and separate territory. For a nation-race to continue to exist, it had to create a unique national culture.

> A few years ago, I asked myself: . . . Where does the deep feeling of national self-identity originate? . . . [I]t is clear that the source of the national feeling should not be sought in education but rather in something that precedes education. In what? I studied this question in depth and answered: in the blood. . . . The feeling of national-identity is ingrained in the man's "blood," in his physical-racial type, and in it only. We do not believe that the spirit is independent of the body; we believe that man's temperament is dependent, first of all on his physical structure. No kind of education, not from his family, and not even from the environment, will make a man who is by nature calm, excitable and hasty. The psychic structure of the nation reflects the physical type even more fully and completely than the temperament of the individual.[16]

Jabotinsky emphasized the need for the cultural isolation of the nation as a precondition for it becoming humanly productive. In turn, cultural isolation required physical isolation. In modern terms, the physical and cultural identities must coevolve. In "A Lecture on Jewish History" (1933) he asserted: "Each race with clear characteristics aspires to become a 'nation,' that is to say, to create for itself a unique economic, political, and spiritual environment."[17] That the Jews have comprised a distinct immanent race explained much of their role in history and suggested that as a race they could still contribute significantly to the future of humanity. The Zionist ideal should be conceived not solely as a way of benefiting the Jews, but also as a contribution to human progress.

Notions of race pervaded the thinking of many Jews and non-Jews well into the late 1930s, by which time *völkisch* conceptions were firmly established

16. Ze'ev Vladimir Jabotinsky, "The Race," quoted in Raphaella Bilski Ben-Hur, *Every Individual, a King: The Social and Political Thought of Ze'ev Vladimir Jabotinsky* (Washington: B'nai B'rith Books, 1993), 91.

17. Ze'ev Vladimir Jabotinsky, "A Lecture on Jewish History," quoted in Ben-Hur, *Every Individual, a King*, 95.

among Zionist intellectuals. For example, at a press conference at the Hebrew University in 1934, the poet Chaim Nachman Bialik (1873–1934) bluntly asserted his Jewish identity: "I too, like Hitler, believe in the power of the blood idea." Jewish willpower and Jewish blood were responsible for undermining what Bialik called "the remnants of paganism" in the Christian world.[18] Although Zionism after World War I tended to be pragmatic and compromising, it retained its ideology. Buber continued to nurture a *völkisch* conception, based on the assumption that in endorsing Jewish nationalism Zionists would lead the world away from chauvinism and towards a humanistic future. However, he was becoming increasingly marginalized, whereas Jabotinsky and his followers progressively promoted an uncompromisingly nationalistic line. The establishment of the State of Israel only intensified the grip of essentialist traditions, rather than adopting the universalistic elements of nineteenth-century Zionism.

A Jewish Race

In the nineteenth century, and increasingly during the first decades of the twentieth, anthropologists conceived of the Jews as a distinct race. Although some emphasized the successful isolation of the Jewish race, others identified Jewish characteristics in the most distant parts of the world.[19] Anthropologists have often presented Jews and gentiles as, respectively, experimental and control populations, arguing that since both communities live under similar conditions, all differences between them are hereditary.[20]

At the beginning of the nineteenth century Zionists who were trained in anthropology (and somewhat later also in genetics) joined the consensus that the Jewish people comprise a race whose specific traits reflect action of selective pressures. For example, the Zionist doctor Aron Sandler, writing in a 1904

18. Chaim Nachman Bialik, "The Present Hour," *Young Zionist* (May 1934): 6–7.

19. This urge to discover "biological" traces of genuine Jewish people (or the Ten Lost Tribes) has also gained currency in some contemporary Orthodox Jewish circles. See Yair Sheleg, "All Wish to Become Jews," *Haaretz*, 17 September 1999, p. 7b. Members of native communities from North Burma, Peru, and elsewhere were brought to Israel under such pretexts. See Raphael Falk, "The Eugenic Dimension of the Settlement of Palestine" (in Hebrew), *Alpayim* 23 (2002): 179–98.

20. John M. Efron, *Defenders of the Race: Jewish Doctors and Race Science in Fin-de-Siècle Europe* (New Haven: Yale University Press, 1994); Mitchell B. Hart, *Social Science and the Politics of Modern Jewish Identity* (Stanford: Stanford University Press, 2000). See also primary sources such as I. M. Judt, *Die Juden als Rasse: Eine Analyse aus dem Gebiete der Anthropologie* (Berlin: Jüdischer Verlag, 1903); Karl Heinrich Stratz, *Was Sind Juden? Eine ethnographisch-anthropologische Studie* (Vienna: Tempsky, 1903).

pamphlet on *Anthropologie und Zionismus*, noted that it was difficult to assert that Jews comprised a pure race, although their mating history indicated that *empirically* they should be considered a race that acquired adaptive traits that in time became inherited. However, Sandler emphatically rejected the claim by some anthropologists that "conservativism" and "liberalism" were also inherited characteristics. He objected that contemporary Zionists displayed the heredity characteristics of the conservative, separationist national zealots, while assimilationists reflected the temperate, liberal universalist views found among the factions of Judea towards the end of the Second Temple era. Furthermore, Sandler rejected the claim that a Zionist homeland would increase whatever deleterious physical and mental qualities of the Jews had been produced by inbreeding over many generations. Instead he argued that Jewish history showed that inbreeding need not always be detrimental and that Zionism would actually decrease inbreeding among the Jewish people by gathering Jews from diverse communities all over the world.[21]

For the sociologist Arthur Ruppin (1876–1943), Zionism was an outgrowth of his activist, universal, and humane *Weltanschauung*. Ruppin had studied physical anthropology with one of its founders in Germany, Felix von Luschan. In 1899, during his first term in Berlin, he became interested in social Darwinism and entered a Krupp prize essay competition that posed the question: "What do the principles of the theory of evolution teach us about the internal political development and legislation of the nations?" Although the first prize went to Wilhelm Schallmayer, whose essay *Vererbung und Auslese* (Heredity and Selection) heralded the birth of German *Rassenhygiene*, Ruppin's essay on *Darwinismus und Sozialwissenschaft* (Darwinism and Social Science) was awarded the second prize.[22] During 1902–3 he acquainted himself with Jewish life in Europe and came into contact with a circle of "practical Zionists" that gathered round Martin Buber, who "were dreaming of Jewish settlements in Palestine."[23] Ruppin's *Die Juden der Gegenwart* (1904, The Jews of Today) contained an anthropological analysis of the Jewish people. His conclusion, which was based on racial-biological, historical-cultural, and religious arguments, portrayed contemporary Jews as comprising a national entity. Ruppin subsequently traveled to Palestine and settled there in 1908 as head of the Palestine Office of the Palestine Land Development Company. In his well-known *Soziologie der*

21. Aron Sandler, *Anthropologie und Zionismus* (Brünn: Jüdischer Buch-und Kunstverlag, 1904).
22. Arthur Ruppin, *Darwinismus und Sozialwissenschaft* (Jena: Fischer, 1903).
23. Alex Bein, *Arthur Ruppin: Memoirs, Diaries, Letters* (London: Weidenfeld and Nicolson, 1971), 75.

Juden (1931) he merely reiterated what appeared to him to be well-established claims about the race and racial characteristics of the Jews:

> I joined the Zionist movement under the slogan "against political Zionism (i.e., Herzl's idea of a charter) and for practical work in Palestine." I wanted to base the right of the Jews to come to Palestine not on some "political" agreement and concession, but on their historical and racial connection to Palestine, and I wanted them to earn the rights in Palestine for themselves through their work there.[24]

Drawing the analogy between the benefits of diversity within a human community and diversity among races, he asserted that it was in the interest of the human race to maintain Jews as a distinct biological entity (as much as a culturally distinct group), rather than to allow them to disappear through assimilation.

> Have the Jews a right to a separate existence? The very question is an insult to the Jewish people, since no other people is required to defend by argument its right to survive. . . . The Jews might justly claim that a history extending over 3,000 years is sufficient justification of their continued existence. . . . But these, after all, are only sentimental arguments; . . . we must first answer the following question: "Can the Jews do more for humanity by remaining a separate nationality than by becoming absorbed in other nations?" A people can be of use to humanity in two ways, firstly through its race-value, i.e., through the spiritual and mental powers incorporated in it, and secondly through its culture. Whoever defends the right of the Jews to a separate existence must do so either in view of their racial or of their cultural value.[25]

Ruppin remained true to his anthropologist's view of the Jews as a race, even in 1938 when he corresponded with the racial anthropologist Hans F. K. Günther, an enthusiastic supporter of the Nazi regime.

Ruppin rejected the claim that intermarriage with the gentile population would result in the Jews becoming physically impaired. Such fears, he claimed, were unfounded because the two populations were very similar in their anthropological characteristics. As for the diverse Jewish communities,

> [t]heir "purity," must be settled as follows: From the moment they left Palestine the Jews absorbed blood of many different non-Jewish peoples. However, racially these people were primarily of the same three racial components from which the Jews of Palestine were earlier established. This is the reason why the majority of the recent Jews are similar in their racial construction to their ancient forefathers in the Land of Israel.[26]

24. Ibid.
25. Arthur Ruppin, *The Jews of To-Day* (New York: Holt, 1913), 212–13.
26. Arthur Ruppin, *Soziologie der Juden* (Berlin: Jüdischer Verlag, 1931), 30.

After having lived for many years in Palestine, where he encountered diverse types of Jews from all over the world, Ruppin's attention increasingly shifted to studying diversity among the Jewish communities. In the 1930s he accepted that the original gene pool of the ancient Israelites was not at all unique, but instead contained a mixture derived from local tribes. The so-called common Jewish racial genetic elements reflected the genes of the nations with which they came in contact in their wandering over the centuries. Furthermore, the "original" gene pool of the Jews in their historical homeland was not a pure racial type.

A similar early Zionist response to crude racism that presented the Jewish national idea in an anthropological liberal-universal light was that of the Czech-Jewish physician and anthropologist Ignaz Zollschan (1877–1948). According to him a "pure Jewish race" knew how to safeguard its inherent cultural abilities (*Kulturfähigkeit*). Consequently, the cosmopolitan importance of the Jewish people lay in its cultural singularity. Being such a highly bred race (*hochgezüchtete Rasse*), it required territorial autonomy in order to guarantee its regeneration.[27] Like Ruppin, Zollschan "espoused Zionism in the belief that by recognizing their racial identity Jews could be redeemed from the degenerative contingencies of the Diaspora," and could restore self-respect and ethnic homogeneity. In their new homeland "the racial individuality and culture of the Jews would be safeguarded."[28] Zollschan spent two years (1927–28) in Palestine. However, unlike Ruppin, who sought to work alongside those who advocated German race regeneration, from the outset Zollschan adopted a combative tone on the subject of race. As Weindling shows in chapter 6 of this volume, Zollschan subsequently radically revised his scientific ideas of Jewish race purity and became committed to an international coalition against Nazi racism.

Eugenics and the Jews

In 1910 the *Jewish Chronicle* published an interview with the aging eugenicist Francis Galton (1822–1911) on "Eugenics and the Jew." The writer praised Galton, "who devoted a long life to the pursuance of a high ideal—that of improving the fitness of the human race and striving to secure that the children born into the world shall be well born," pointing out "that from the days of Moses Jews have been 'eugenists,' apart from the hygienic laws enjoined in the Mosaic code, which affect the individual rather than the race." Asked

27. Ignaz Zollschan, *Das Rassenproblem: Unter besonderer Berüksichtigung der theoretischen Grundlagen der jüdischen Rassenfrage*, 2nd ed. (Vienna: Braumüller, 1911), 426–27.
28. Paul Weindling, chapter 6 of this volume, 122.

what effect persecution had on the Jewish race, Galton answered: "So far as persecution weeds out those who are unfit so far it tends to evolve a race suited to meet the conditions." Asked whether it was immoral to view persecution as a positive aid to improving race culture, he exhorted: "It is not immoral but unmoral—it has nothing to do with morals. . . . It is the aim of eugenics to supply many means by which the effects of these drastic and not always successful aids to race culture may be produced in a more scientific and kindly way."[29]

Early in the twentieth century the status of eugenics was enhanced by the establishment of the new science of genetics. Already in the first volume of the *Journal of Genetics* of 1911, the physician, virologist, and enthusiastic eugenicist Redcliffe Nathan Salaman (1874–1955) published a paper in which he traced the progeny of Jewish-gentile marriages and those of their offspring who married Jews. He concluded that "the Jewish facial type . . . is a character which is subject to the Mendelian law of Heredity," and that Jewish facial features differed from gentile features essentially by a recessive allele of a single gene.[30] During World War I, Salaman served in the British army as a doctor for the Jewish battalions in Egypt and Palestine. Impressed by the physique of the sons of the settlers in the Judean Zionist colonies—blonde-haired youngsters who were an equal match for the British soldiers at sports—he concluded that he was witnessing natural selection in action. The Zionist settlers who lived physically demanding outdoor lives as farmers favored the Philistine genetic component of the ancient Jewish gene pool, as manifested by their fair hair, which had been suppressed when Jews lived degenerate lives in the Diaspora.[31]

Although Rudolf Virchow's 1886 study of German schoolchildren contained the surprising result that Jewish children were almost as frequently fair-haired as their Aryan counterparts, the issue of fair hair among the Jews resurfaced repeatedly. In 1892 the German anthropologist Felix von Luschan proposed the hypothesis that red-haired Jews reflected the Amorite element of the ancient Israelites. However, in 1903 the American Jewish demographer Maurice Fishberg examined similar evidence but rejected Luschan's conjecture. The subject arose again in 1935 when a physician from Lemberg (Lvov) wrote "On the Origin of the Yellowish Jews" in the Hebrew journal of the Medical Association in Palestine: "The color of the skin, hair, and eyes is a

29. Francis Galton, "Eugenics and the Jew," *Jewish Chronicle*, 29 July 1910, 16.

30. Redcliffe N. Salaman, "Heredity and the Jew," *Journal of Genetics* 1 (1911): 273–92.

31. Redcliffe N. Salaman, "What Has Become of the Philistines?," *Quarterly Statement of Palestine Exploration Fund* 57 (1925): 1–17.

prominent marker of race. [However,] a conspicuous fact is that the distribution of fair types among Jews is by and large correlated to that of the peoples among whom they live." Based on his research he accepted Fishberg's conclusion with respect to the fair-skinned Jews as the sounder one.[32]

Interestingly, Salaman barely discussed "Jewish diseases," although it was known that Jews were afflicted by typical hereditary illnesses. Mental diseases and the psychoses of the Jews were favored subjects for disputes about the adverse effects of selection on the biology of the Jews and the impact of inbreeding in compounding these characteristics. For example, in a lecture on "Illness and Rates of Death among the Jews," delivered in March 1905 at the Verein für Statistik der Juden in Munich, Dr. M. Epstein argued that the higher rates of mental illness found among Jews can be explained only biologically. This propensity was not a social product, but rather an "inherited neuropathic trait" that has been intensified through inbreeding as well as through the nervousness of the Jew. "Their nervousness is a radical trait. It is a product of the selection process," he claimed.[33]

One dissenting voice was that of Shneor Zalman Bychowski (1865–1934), a Zionist doctor from Warsaw. In 1918 he published in the Hebrew literary magazine *Ha-Tekufah* an article on "Nervous Disease and the Eugenics of the Jews."[34] Bychowski pointed out that "among the experts in neuropathology it is considered as settled that Jews are more liable to suffer from nervous diseases. However, this opinion is lacking a solid foundation, and when we examine it, we'll find that it needs to be tested, and it should be thoroughly reconsidered and questioned."[35] He referred to Jean-Martin Charcot (1825–93)—with whom Sigmund Freud had studied—who was consulted by many Polish and Lithuanian Jews. Charcot defined their suffering as that of "the Wandering Jew" (*le juif errant*).[36] According to Bychowski, Charcot recognized that these wretched Jews did not appear to suffer from any specific clinical illness. Yet they had already attended almost every famous doctor in Europe. Although their complaints extended over many years, they never

32. S. Tshurtakover, "On the Origin of the Yellowish Jews" (in Hebrew), *Ha-Refu'ah* 14 (1936): 106–10.

33. Cited in Hart, *Social Science*, 271.

34. Raphael Falk, "Nervous Diseases and Eugenics of the Jews: A View from 1918," *Korot* 17 (2003–4): 23–46.

35. Shneor Zalman Bychowski, "Nervous Disease and the Eugenics of the Jews" (in Hebrew), *Ha-Tekufah* 2 (1918): 289–307, on 289.

36. On the "Wandering Jew" see Jan Goldstein, "The Wandering Jew and the Problem of Psychiatric Anti-Semitism in Fin-de-Siècle France," *Journal of Contemporary History* 20 (1985): 521–52.

systematically took any medicine. Each of them had accumulated many prescriptions and doctors' orders, which they kept carefully, although never following any doctor's advice.

Examining the epidemiological evidence Bychowski rejected most of the arguments of his predecessors who had identified either inherited factors or certain environmental causes. Instead he suggested that "perigenic causes"—essentially the effects of the appalling living conditions and persecutions that Russian and Lithuanian Jews had suffered—were responsible for "Jewish" nervous afflictions.

> From the eugenic perspective there is no reason to fear for the future. As I have shown, the impressions that the perigenic factors make . . . are temporary and may be erased. The nervous potential of the Jews has remained whole, with no permanent defect.[37]

Bychowski did not, however, reject eugenics—quite the opposite. Thus, he deplored the habit among Polish and Lithuanian Jews of not permitting a man to remain single even when he was sick and might transmit his disease to his progeny. He also recommended harsh eugenic measures, especially to

> those who build the future of the nation—the Zionists. The resurrection of the nation in its homeland will be possible only if the "human material" that goes there is healthy. In this respect it will be necessary to employ from the beginning strict means, like the "law" against immigration that has been instituted in the United States. It is of special significance that the Zionists should learn to view marriage not as a personal act that a person may handle as his heart may desire, but rather as an important public act, on which depends the future of the race. . . . If there are reasons to believe that the marriage may produce sick children, these must be strictly forbidden. The Zionists must be especially careful, when they come to rebuild anew the life of the nation.[38]

Eugenic awareness clearly formed part of the Zionists' call for Jews to resume "normal" lifestyles.

Eugenics in Zion

The years between the two World Wars witnessed increasing Zionist emigration to Palestine, or, as it was called, the Land of Israel. This was accompanied by a rising consciousness about the eugenic aspects of Jewish life, especially settlement in the Land of Israel. Although the scientific community

37. Bychowski, "Nervous Disease," 306.
38. Ibid, 299.

became increasingly disenchanted with eugenics, which was generally considered oversimplistic and too deterministic, its sociopolitical implications thrived. Among ideologically motivated Zionist settlers, several leading physicians and educators became the flag bearers for the campaign to promote eugenic aspects of Zionism.

Indicative of the climate of the time was the "call to doctors and teachers" by the educator and literary critic Israel Rubin, published in the official organ of the Hebrew Authors' Association, entitled "The Ingathering of the Diasporas from a Eugenic Perspective." Rubin explicitly viewed "our life in the homeland, in its very essence, to be primarily a great and courageous national effort in the eugenic sense."

> Anyone, who does not recognize in the return of the sons to the land of their forefathers a great *eugenic* revolution in the life of the nation, does not discern the "forest" from the individual trees. . . . The essence is the sum total: *The production of a New Hebrew type* restored and improved. Thus, a *psychobiological approach to the problem of the settlement of the Land of Israel* is a duty of us all![39]

As early as 1922 Mordechai Bruchov, who had served as physician to the Herzelia gymnasium, the first Hebrew high school in Palestine, and later headed the Department of Hygiene at the Hadassah Hospital, published an article emphasizing that knowledge of heredity is required for "the life of both the individual and the state." In a society of settlers "the prevailing spirit is the idea that the greatest sin that people can commit to the god of life is to procreate sick children. . . . In the struggle of nations, in the clandestine 'cultural' struggle of one nation with another, the one wins who provides for the improvement of the race, to the benefit of the biological value of the progeny."[40]

Another physician, Abraham Matmon, in discussing "the improvement of the human species and its meaning to our nation," called attention to the yearly increase in the number of defective people:

> Those at the bottom [of the social ladder] are the ones producing more children and inherit to them their characteristics. And let us not forget the excessive expenses that any nation of culture spends on those degenerate persons. . . . We must take the fate of those persons into our own hands, provide them with the help and shelter they need, and at the same time affect their reproduction and guide it in the proper way. This is the task of modern hygiene: To protect humanity from the flood of inferiors and block the way for them from penetrating

39. Israel Rubin, "Gathering of the Exiles from a Eugenic Perspective" (in Hebrew), *Moznayim* 1 (1934): 89–93. Emphasis in original.
40. Mordechai Bruchov, "Mendelism" (in Hebrew), *Ha-Tekufah* 16 (1922): 326–41.

> humanity, by denying them the possibility to inherit their delinquency to later generations.[41]

Reviewing the subject of "Heredity and Environment," the educator Nissan Touroff agreed that it was reasonable to assume that both the physical and mental characteristics of Diaspora Jews are the outcomes of natural selection. He, however, called for the preservation of these mental characteristics:

> The excessive unfolding of mental talents is prevalent among children of Israel nearly everywhere. . . . Thus there is ground to claim that our natural inheritance—whatever its biological and historical causes may be—is far from being dire.
>
> If indeed, with time, external conditions will change as much as to guarantee us free existence and a possibility for truthful self-expression of our hereditary sources—we may be able again to participate proudly in the symphony of general human creation and our honor will be repaired.
>
> Is there any need to name the only corner on Earth where this dream may be realized?[42]

Analyzing the parental guidance articles and books that were published in Palestine from the 1920s onwards Sachlav Stoler-Liss has pointed out that the family was considered not only as a means of reproduction—a high-priority Zionist cause—but mainly as an instrument for radically changing the national image. The dominant theme was "A Sound Mind in a Sound Body."[43] Stoler-Liss concluded that in contrast to the 1930s, when the emphasis was on the purity of the race and the quality of children required to improve the nation, the emphasis subsequently shifted to the need to increase the birthrate in order to catch up with the high birthrate of the neighboring nations. Moreover, there has been a long history of institutional intervention in childcare that continues today: "Child care is conceived as part of a national project, and is always measured in relation to army service." Stoler-Liss was careful to add: "This is not a ritual in the spirit of Fascist

41. Abraham Matmon, *The Improvement of the Race of the Human Species and Its Meaning to Our Nation* (in Hebrew) (Tel-Aviv: Biological-Hygienic Library, 1933), 5–6.

42. Nissan Touroff, "Heredity and Environment" (in Hebrew), *Ha-Ḥinukh* 11 (1938): 274–92, on 292.

43. Sachlav Stoler-Liss, "Zionist Baby and Child Care: Anthropological Analysis of Parent's Manuals," master's thesis, Tel-Aviv University, 1998, points out that the demand for "A Sound Mind in a Sound Body," whether resulting from national and racist interests or from a wish to improve the world aesthetically, is accompanied by a vicious hatred of those who are different—handicapped either in body or mind.

states. Yet, every mother who raised her child in Israel, in the past and at present, is conscious that this is not only her personal task, but rather a national task the climax of which—at the age of eighteen—is the recruitment of the Zionist baby to the nation's army." She emphasized, however, that "just as it is impossible to explain the growth of the Zionist movement in Europe out of the context of the growth of nationalism, one cannot understand the Zionist ideology concerning the family without reference to the attitude of the European national movements toward the family."[44]

To illustrate the impact of eugenics on Zionist family life Stoler-Liss cited the opening page of the 1934 annual parents' guide *For Mother and Child* written by Joseph Mayer, the chief executive of the Labor Organization Sick Fund, the major health insurance fund for Jews in Palestine:

> Who has the right to give birth to children? Eugenics, the science for the improvement of the race and keeping it from degeneration, is concerned with searching for a proper answer to this question. . . . Cases of [marriages of carriers of hereditary diseases] are not rare in all nations, and especially in the Hebrew nation, which lived in exile for eighteen centuries. Now our nation is resurrected to life in nature in the homeland. Isn't it our duty to see to it that we would have sons who are healthy in body and in mind? For us "eugenics," and especially the prevention of transfer of hereditary diseases, is of even greater importance than it is for other nations! . . . Doctors, sportsmen, and national public figures must carry on with effective propaganda for the idea: Do not procreate children if you are not sure that they will be healthy in body and mind.[45]

The application of insights from science to Zionist activities was also integral to the program of the professional biologists. Fritz S. Bodenheimer (1897–1959) was the son of one of Herzl's closest allies. During World War I, after encountering anti-Semitism in the German army and meeting Jews from Eastern Europe, he adopted Zionism and studied agricultural entomology, which he intended to apply to the Zionist cause. He arrived in Palestine in 1922 and joined the Hebrew University in Jerusalem in 1928. In the preface to his *Biological Foundations of Populations Theory* (1936) he stated that he had written it in order to call the attention of Jewish settlers to the biological condition of humankind in general, and especially its role in national politics. The last chapter of the book was devoted to "Race, Inheritance, and Eugenics of Humans":

> The term "race" in humans is that of a type similar in its physical and mental foundations. Among contemporary Jews we find sometimes pure types of

44. Interview with Adi Katz, *Sofshavua*, 19 May 2000, 70–76.
45. Stoler-Liss, "Zionist Baby," 3.

Semite—the Hittite, or the Philistine of ancient times—as depicted on the ancient Assyrian and Egyptian sculptures . . . . On the other hand, one finds among contemporary Jews some other types—Slavs, Tatars, etc. This indicates the complex composition of different elements that merged into the type called today "the Jewish race." Every nation of historical significance is formed by a huge mixing of several different races.[46]

For Bodenheimer, as for most settlers, the diversity of Jewish communities became a central issue of Jewish identity. He stressed the importance of a common biological—i.e., Jewish—element in guaranteeing the harmonious vigor that is produced by interethnic marriages. However, such intermarriages were also important for the Zionist political cause because they surmount both the threat of social disparity within the community and the external threat posed by the faster reproductive rate of the indigenous Arab population. Nationalism clearly underpinned Bodenheimer's operational conclusions pertaining to the "racial purity" of Jewish communities.

It may be assumed with certainty that elements of the Jewish race that we call the "Sephardi" had not participated in the great mixture with the Slavic and Tataric nations that profoundly affected the Ashkenazi during the Khazars' times and in the days of Chmielnicki.[47] On the other hand, we do not find among the Ashkenazi the mixture with the black types, at least not to the great extent found among the Sephardi. . . . Zionism aspires directly . . . to the unification of all the elements prevalent today among the Jewish race, in order to form a new harmonious Jewish type.

In recent times we find in the professional literature the opinion that racial mixture is not desired for the development of nations. Undoubtedly, this opinion only reflects contemporary politics. . . .

Concerning the situation in our settlements in Palestine, it is the lower classes, which are on a low cultural level, that are the most fertile among us. [Notwithstanding,] in this case it is of great virtue, because the Yishuv[48] is in great danger from the high reproductive rate among the Arabs. . . . Thus, it is important for the Yishuv that there are among us elements who have small needs and know how to live under harsh conditions, that are rather outstanding in fertility.[49]

46. Fritz S. Bodenheimer, *The Biological Foundations of Populations Theory* (in Hebrew) (Tel-Aviv: Stybel, 1936), 137–38.

47. The riots during the Cossack revolt led by Bogdan Chmielnicki (1648–49).

48. Hebrew for "settlement," the community of Jewish settlers in Palestine.

49. Bodenheimer, *Biological Foundations*, 140–42.

Such Zionist settlers not only conceived their national task in political or cultural terms, but they also saw a need to define the Jewish nation as an anthropological entity that must make use of eugenics to return to "normality."

During the 1920s and '30s, when the immigrants were almost exclusively from European countries, a debate raged between supporters of the Zionist tenet of unrestricted immigration for all Jews and those calling for selection based on guidelines that included medical criteria. The ideal of building a vigorous and healthy nation after two millennia of degenerative living in the Diaspora heavily influenced the absorption policy of the Zionist Organization, and eugenics considerations significantly influenced its settlement policy.[50]

In 1919, Ruppin, then head of the Palestine Office of the Zionist Federation, presented the position of the Zionist leadership and established the future practice for "choosing the human material":

> One of the most important questions we face in the settlement in Palestine is the choice of the human material. . . . To date . . . every Jew who immigrated to Palestine was gladly received; and even if he was old, sick, unqualified for work, or antisocial in character; public opinion in Palestine didn't ask questions about this. . . . it would be better if only healthy people with all their needs and their powers would come to Palestine so new generations will arise in the country that are healthy and strong.[51]

In his conclusion Ruppin made fairly explicit the link between race and eugenics. He wondered "whether it is not possible to have an impact in the direction of purifying the Jewish race" and stressed that "in Palestine we want to develop particularly what is Jewish."

As early as the 1920s the Zionist Federation set up a network of physicians who carried out medical inspections of immigration candidates, both in the immigrants' country of origin and after they had arrived in Palestine. If a young immigrant was found to suffer from an illness that might adversely affect his or her chances of economic or social integration in the Yishuv, he or she was returned to the country of origin, a step coordinated with the returnee's family.[52] In recent years, however, critics have charged that the restrictions based on medical grounds, operated by Israeli immigration

50. Nadav Davidovitch and Shifra Shvarts, "Health, Zionism, and Ideology: Medical Selection of Jewish European Immigrants to Palestine," in *Facing Illness in Troubled Times: Health in Europe in the Interwar Years, 1918–1939*, ed. Iris Borowy and Wolf D. Gruner (Berlin: Lang, 2005), 409–24.

51. Arthur Ruppin, "Die Auslese des Menschenmaterials für Palästina," *Der Jude* 3 (1918–19): 373–83.

52. Davidovitch and Shvarts, "Health, Zionism, and Ideology."

authorities in the 1950s and '60s, concealed an underlying ethnic bias intended to reduce the number of Sephardi and Oriental Jews.

Genetics of the Jews

Matters changed radically in the wake of the Holocaust and World War II. Eugenics became taboo and scientific discussions about race were transformed.[53] Population genetics now largely replaced anthropological types in tackling issues of race, as becomes clear if we compare the definition of race given by Bodenheimer in 1936 (cited above) and that used in 1952 by the Swedish demographer and population geneticist Gunnar Dahlberg:

> Particularly some years ago, when Sweden too had its quota of Nazi sympathizers, it was often asked whether the Jews really were a race. The question frequently came from members of the so-called socialist intelligentsia who detested anti-Semitism and required an authoritative reassurance that the Jews were not a distinct race.[54]

Dahlberg defined race as "an isolate that differs genetically from other isolates." An isolate may be geographically or socially bounded, but the boundaries need not be sharp. In other words, we appeal to the notion of race in order to identify and stress conspicuous differences between populations.

> The Jews are as a matter of fact by some considered as a racial group, i.e., they make up several races of the second degree whose specific differences are rather insignificant. . . . The bond that has kept the Jews together is obviously their religion; and as a result of that and anti-Semitism the Jews have a different sociological tradition and have been subjected to particularly strict selection, which may have created genetic differences.[55]

In the wake of the Holocaust, Zionist policy was redirected to saving survivors and reconstructing the Jewish people. For a time Zionist agencies eschewed all medical selection. But following the establishment of the State of Israel, priorities shifted to those necessary to secure the state's existence. A fierce debate soon commenced on the application of medical selection criteria to the waves of immigrants entering the country, mostly from North Africa and the Middle East.[56]

53. Elazar Barkan, *The Retreat of Scientific Racism* (Cambridge: Cambridge University Press, 1992).
54. Gunnar Dahlberg, "The Swedish Jews: A Discussion of the Basic Concept," *Acta Genetica et Statistica Medica* 3 (1952): 106–17, on 106.
55. Ibid, 109.
56. Davidovitch and Shvarts, "Health, Zionism, and Ideology."

In Israel the earlier dispute over whether the Jews constituted a distinct race was replaced by the question "Who is a Jew?" Which criteria should be applied in defining who is Jewish? The issues were further complicated by extensive immigration, often of whole communities of people who were considered "esoteric," even from the perspective of the indigenous populations that had earlier enhanced ethnic heterogeneity. This debate also presented geneticists with a unique challenge as better parameters became available for the study of the genetic components of phenotypic variability, thus providing geneticists with the opportunity to trace the dynamics of human isolates and to follow in real time the merging of populations.

In the 1950s genetic studies based on blood-group polymorphisms were carried out in different communities in Israel, primarily by Yosef Gurevitch and his coworkers.[57] An increasing scientific effort was directed to uncovering the common genetic characteristics of the Jews[58] and to trying to establish links with the more "esoteric" communities. In September 1961 the Jerusalem conference on "The Genetics of Migrant and Isolate Populations" emphasized Israel's unique position:

> The study of isolate and migrant populations is valuable to the geneticist because such groups offer a "laboratory" in which certain variables can be observed to operate largely without contamination. . . . [T]he population of Israel offers the geneticist a unique opportunity. There, within the present population, one finds native-born Jews, both isolate and assimilated, urban and rural; immigrant isolate groups from various parts of the world that maintain their isolate characteristics; migrant groups that became isolates after immigration; and isolates that assimilated upon entering Israel. And for all these groups accurate demographic records are available. . . . These applications of the tools of modern genetics to populations of ancient origin may be as valuable for the questions they raise as for the answers they provide, but they contribute significantly to our understanding of many genetic traits in a variety of ethnic groups.[59]

Subsequent research on the genetic polymorphism of Israeli communities was presented at several later conferences.[60]

57. Gurevitch et al. published several papers in the 1950s including "Rh Blood Types in Jerusalem," *Annals of Eugenics, London* 16 (1951): 129–30, and "Blood Groups in Persian Jews," *Annals of Human Genetics* 21 (1956): 135–38.

58. A. E. Mourant, Ada C. Kopec, and Kazimiera Domanniwska-Sobczak, *The Genetics of the Jews* (Oxford: Oxford University Press, 1978).

59. Text on dust jacket of Elizabeth Goldschmidt, *The Genetics of Migrant and Isolate Populations* (New York: Williams and Wilkins, 1963).

60. See Richard M. Goodman, *Genetic Disorders among the Jewish People* (Baltimore: Johns Hopkins University Press, 1979); Richard M. Goodman and Arno G. Motulsky, *Genetic Diseases*

In their study of genetic data on Jewish populations Batsheva Bonné-Tamir, Samuel Karlin, and Ron Kenett wrote:

> Our purpose in studying the differences and similarities between various Jewish populations was not to determine whether a Jewish race exists, nor was it to discover the original genes of "ancient Hebrews," or to retrieve genetic characteristics in the historical development of the Jews. Rather, it was to evaluate the extent of "heterogeneity" in the separate populations, to construct a profile of each population as shaped by the genetic data, and to draw inferences about the possible influences of dispersion, migration, and admixture processes on the genetic composition of these populations.[61]

However, by asserting what was *not* the purpose of this study, the authors suggested a covert need to contribute to the national sociopolitical debate. Implicitly they provided evidence that at least some blood is common to the various Jewish communities.[62]

This ambivalence is perhaps best revealed in the demographer Helmut Muhsam's 1964 effort to present an impartial method for studying the genetic origin of the Jews. To be on the safe side, Muhsam considered only attributes "which are completely determined by one gene and miscegenation will be considered as a process of amalgamation." He plotted on a two-dimensional graph the frequencies of the A and B alleles of the ABO blood types of the Jewish *eidot* (communities) and the corresponding frequencies for the non-Jewish communities among which they lived—their "genetic environment." Muhsam reasoned that if there existed a nuclear basic Jewish gene pool that was dispersed throughout the Diaspora and intermingled to some extent with the non-Jewish "genetic environment," then the extension of vectors from the non-Jewish communities to their corresponding Jewish communities should converge on the values of the ancient common gene pool of the Jews. However, the data from thirty-six communities did not confirm his prediction. Nevertheless, after an extensive discussion Muhsam concluded with an evident sigh of relief:

> It would not require us, in view of the observed data, to abandon the assumption that each *eidah* is a mixture of a "genuine" Jewish group and its [genetic]

among Ashkenazi Jews (New York: Raven Press, 1979); Bracha Ramot, *Genetic Polymorphism and Diseases in Man* (New York: Academic Press, 1974); Batsheva Bonné-Tamir and Avinoam Adam, eds., *Genetic Diversity among the Jews* (New York: Oxford University Press, 1992).

61. B. Bonné-Tamir, S. Karlin, and R. Kenett, "Analysis of Genetic Data on Jewish Populations. I. Historical Background, Demographic Features, and Genetic Markers," *American Journal of Human Genetics* 31 (1979): 324–40, on 325.

62. Raphael Falk, "Zionism and the Biology of the Jews," *Science in Context* 11 (1998): 587–607.

> environment, where the Jewish group is called "genuine" because it stems directly from the original pool.[63]

Of particular interest is the attempt by the physician Chaim Sheba and his associates to reconstruct Jewish history from epidemiological-genetic data.[64] Combining the conventional historical account with the epidemiological picture of endemic, genetically oriented diseases, Sheba traced the history of the Jews from the eighth century BCE onward by using indicators such as the relatively high frequencies of hereditary diseases in various communities.

> Why do European (Ashkenazi) Jews lack or practically lack G6PD deficiency [Dubin Johnson syndrome], phenylketonuria, and the thalassemias? For G6PD deficiency, which is an X-linked mutation, . . . if you read *The Jewish Wars* of Flavius Josephus [38–100 CE], you will learn that the Roman exile into Italy consisted of males only who were sold as slaves or thrown to the lions in the arena. . . . Thus, under the conditions of slavery the Semite males had to marry, and obviously convert, the Japhethite females. Therefore the core of those Jews who moved into Europe . . . did not carry the mutant gene for G6PD deficiency to their male offspring at all.[65]

No alternatives to the stochastic distribution of rare mutations were considered, such as dependence on environmental conditions such as climate or parasitic pathogenic vectors.[66]

With the advent of more precise methods in molecular genetics and especially methods for detecting DNA sequences as markers, the study of genetic polymorphisms and the construction of phylogenies became more authoritative and more amenable to manipulation, i.e., to eugenics. Several hereditary-linked handicaps of inhabitants of the Levant (e.g., thalassemia, familial Mediterranean fever) turned out to be due to numerous mutations at the DNA level. Often, however, the interpretation of the causal path for sequence similarities depended on the preconceptions held by the authors. The intensification of political conflicts further exacerbated these tendencies. Thus, for example, the discovery of common Y-chromosome sequences among Jews from Hadhramaut and among men of the Lemba tribe in Zimbabwe and South Africa was taken to support Lemba claims, based on their cultural

63. Helmut V. Muhsam, "The Genetic Origin of the Jews," *Genus* 20 (1964): 3–30, on 21.
64. Chaim Sheba, "An Attempt to Re-construct the Wanderings of the Children of Israel Using Biochemical Tests" (in Hebrew), *Madda* 4 (1960): 34–39.
65. Chaim Sheba, "Jewish Migration in Its Historical Perspective" (in Hebrew), *Ha-Refu'ah* 81 (1971): 1333–41, on 1338.
66. See Goldschmidt, *The Genetics of Migrant and Isolate Populations*; Arno G. Motulsky, "Jewish Diseases and Origins," *Nature Genetics* 9 (1995): 99–101.

traditions, that they constitute a Jewish community that migrated from southern Arabia. The Habbanite Jewish community from Hadhramaut that emigrated to Israel in 1950 identified its origin in the Land of Israel after the destruction of the First Temple in 586 BCE. Most scholars dismiss the claim that the original Jews of southern Arabia were local people who were converted to Judaism. However, according to some Yemeni scholars Yemenite Jews form an integral part of the Yemenite nation, and their emigration to the Land of Israel was an explicit Zionist colonization, rather than a religious act.[67] The possibility that the Lemba stem from some Arabian (Jewish or non-Jewish) slave merchants is also ignored. Similarly, the discovery that the same mutation of blood clotting factor XI is found in both Ashkenazi and Iraqi Jews (whereas a second mutation was limited to Ashkenazim) was interpreted as indicating the existence of an ancient common gene pool of the Ashkenazi and Oriental Jews, thus supposedly "finally excluding" claims that Ashkenazi are of Khazar stock.[68] The possibility that one or a few carriers of the gene from the Ashkenazi community married a spouse of the Oriental community—such marriages are historically documented—who transmitted the gene has not been discussed.

DNA sequencing (that now encompassed many, rather than single-gene comparisons) boosted the construction of phylogenies of Jewish communities that purportedly existed for thousands of years and helped establish their relatedness to non-Jewish communities—particularly Mediterranean communities—among whom they lived.[69] Two significant patterns emerge from these studies: First, the cluster of data from diverse Jewish communities and the close correlations with the cluster of Arab communities. Second, these two overlapping clusters were clearly distinct from the loose clusters of Europeans, on the one hand, and of Africans, on the other.

Hammer et al. addressed the racial question squarely: "Given the complex history of migration, can Jews be traced to a single Middle Eastern ancestry, or are present-day Jewish communities more closely related to non-Jewish populations from the same geographic area?" They concluded that the results

67. Mohamed Khatam El-Qudai and Mohamed Ben-Salem, in the *Yemen Times*, quoted in *Haaretz*, 15 October 1999, p. B8.

68. Hava Peretz et al., "The Two Common Mutations Causing Factor XI Deficiency in Jews Stem from Distinct Founders," *Blood* 90 (1997): 2654–59; Uri Seligsohn, "Genetic and Historic Aspects of the Blood Clotting Systems in Jews" (in Hebrew), *Igeret, The Israeli National Academy of Science*, no. 22, May 2002, 7–8, on 8.

69. Almut Nebel et al., "High-Resolution Y Chromosome Haplotypes of Israeli and Palestinian Arabs Reveal Geographic Substructure and Substantial Overlap with Haplotypes of Jews," *Human Genetics* 107 (2000): 630–41.

suggest that common ancestry is the major determinant of the genetic variability observed among Jewish communities, with admixture playing a secondary role.[70] Similarly, Amar et al. introduced their discussion with the statement: "The genetic makeup of today's Jewish population is the product of the common ancestral gene pool and the introduction from the peoples among whom, over the ages, the Jews lived." They claim that their study has confirmed the hypothesis "that Jews share common features, a fact that points to a common ancestry. A certain degree of admixture with their pre-immigration neighbors exists despite the cultural and religious constraints against intermarriage."[71]

The possibility that those Palestinian Arabs who have lived in the country for centuries may be the progeny of the original inhabitants of biblical times had been suggested in the past, even by some Zionists. Thus in 1917 David Ben-Gurion asserted that "the farmer settlements that the Arab conquerors found in the Land of Israel in the seventh century were none but those of Hebrew settlers that remained in the country."[72] The recent publication in *Human Immunology* of a study by Arnaiz-Villena et al., a Spanish-Palestinian team, on "The Origin of Palestinians and Their Genetic Relatedness with Other Mediterranean Populations" has reopened this contentious topic.[73] They reached essentially the same empirical conclusion as that of the Israeli teams: "Palestinians are generally very close to Jews and other Middle East populations. . . . Thus, Palestinian-Jewish rivalry is based in cultural and religious, but not in genetic differences." The authors, however, presented the historic evidence from a different perspective, emphasizing the Canaanite and Philistine contribution to the ancient inhabitants of the country. Unfortunately, the authors made some overt political comments that led the editors to remove the paper from the journal's Internet version. The incident, however, exposed the profound involvement of so-called neutral scientific discourse on genetic origins in the Palestinian-Israeli political struggle. This was most apparent in a letter from Mazin Qumsiyeh, a

70. M. F. Hammer et al., "Jewish and Middle Eastern Non-Jewish Populations Share a Common Pool of Y-Chromosome Biallelic Haplotype," *Proceedings of the National Academy of Science, Washington* 97 (2000): 6769–74.

71. A. Amar et al., "Molecular Analysis of Hla Class II Polymorphisms among Different Ethnic Groups in Israel," *Human Immunology* 60 (1999): 723–30.

72. David Ben-Gurion, "Concerning the Origins of the *Fellahin*," in Israel Belkind, *The Arabs in the Land of Israel* (in Hebrew) (Tel-Aviv: Hermon, 1969), 43–48.

73. Antonio Arnaiz-Villena et al., "The Origin of Palestinians and Their Genetic Relatedness with Other Mediterranean Populations," *Human Immunology* 62 (2001): 889–900.

Palestinian-American scientist. Qumsiyeh asserted that the data provided by Arnaiz-Villena and colleagues

> is ironically consistent with the data published in the same journal by Israeli scientists. . . . Amar et al. showed that "Israeli Arabs" (Palestinians who are Israeli citizens) are closer to Sephardic Jews than either is to Ashkenazi Jews. Yet, Amar et al. incredibly concluded that "We have shown that Jews share common features, a fact that points to a common ancestry." . . . Many worked feverishly to establish links (however tenuous) between Ashkenazi Jews and the ancient Israelites. . . . But Ashkenazim are also clearly closer to Turkic/Slavic than either Sepharadim or Arab populations.[74]

Qumsiyeh suggested that the data might also be consistent with the hypothesis that Ashkenazi Jews were descended from southeast European and Turkish peoples (the Khazars), whereas the Sephardi Jews are of Arabic descent. He challenged Amar et al.'s statement that they have shown that "the Jews share common features, a fact that points to a common ancestry," whereas according to him "the correct statement from their own data is that some Jews (Sephardim) are more similar to Palestinians than either group is to other Jews (Ashkenazim and Ethiopian Jews)." In other words, the data do not justify the Zionist claim of a single Jewish racial group that can trace its origins to the Israelite population of biblical Land of Israel. Zionism and race are as intertwined today as they were a century ago.

74. "Letter from Dr. Mazin Qumsiyeh to the Society of Histocompatibility and Immunology," *The Ambassadors Magazine* 5 (1), January 2002 (http://ambassadors.net/archives/issue11/opinions2.htm, accessed 15 August 2005).

PART THREE

Evolution and Contemporary Judaism

It is remarkable that, despite the fact that the *Origin of Species* was published almost 150 years ago, evolution remains a controversial issue in many religious circles. Indeed, partly owing to the strength of creationism and the Intelligent Design movement, particularly among evangelical Christians, the subject is considerably more fraught with religious objections than it was fifty or a hundred years ago. Although many Jews may wish to ignore the contemporary furor, an increasing number of writers, thinkers, and educators within the Jewish community are confronting the issues. In this final set of papers we examine four of the many ways in which contemporary Jews are responding.

In the first pair of papers Shai Cherry and Rena Selya analyze how the contemporary Modern Orthodox movement is rising to the challenge of evolution with a combination of both resistive and accommodative strategies. Cherry critically examines the popular writings of three Jewish physicists who advocate approaches for Orthodox Jews to bridge traditional Judaism and science, including evolution. He argues that while these writers adopt some of the strategies used by fundamentalist Christians in opposing evolution, they also try to attract recruits to Modern Orthodoxy by showing that it can embrace evolution, albeit on its own terms. Since the classroom has been a primary site for controversy over evolution, Selya examines the curriculum of Modern Orthodox day schools in order to determine how Orthodox Jews have addressed the teaching of evolution. In contrast to the insistence by many evangelical Christians that creationism and Intelligent Design should be taught alongside evolution, most of the Jewish schools surveyed experience little problem with teaching evolution, although often requiring pupils to explore with a rabbi some of the religious issues raised by the theory.

The reflections of Cherry and Selya on the current Modern Orthodox responses to evolution raise a host of questions. Judging by the relatively small number of articles in the Jewish press, evolution is not a particularly contentious topic for most Jews. However, this rather simplistic assessment needs to be tested by empirical evidence obtained from different sections of the Jewish community throughout the world. What do Jews know about Darwin's theory and the current controversies over Intelligent Design? Do Jews consider Darwin's ideas dangerous because they engender anti-Semitism? Is Darwinism perceived as somehow linked to eugenics? Is evolution considered compatible or incompatible with Judaism and, if the latter, where precisely does the discord lie? The answers to such questions will shed considerable light on the views and concerns of world Jewry. Such studies need to be supplemented by a close analytical reading of the currently available sources, such as the pronouncements by influential rabbis and the circulation of opinions in books, the press, and, increasingly, the Internet. Following Selya's lead, further studies are required to determine how Jewish students encounter evolution in a variety of educational institutions, particularly in Israel where education provisions reflect deep political and religious divisions. The collected evidence will enable researchers to compare and contrast different sections of the Jewish community and across different geographical areas. But comparisons should not end there, because some of the most informative and culturally significant results depend on comparisons between Jewish views and attitudes towards evolution and those of their non-Jewish neighbors, including Christians and Muslims.

In the two final chapters Carl Feit and Lawrence Troster offer different perspectives on the relationship of evolution and Judaism. Feit, a Modern Orthodox rabbi and professor of biology at Yeshiva University, argues that a fairly traditional reading of medieval Jewish texts helps us to understand why evolutionary theory was accepted by Rabbis Abraham Isaac Kook and Joseph B. Soloveitchik, two towering intellects in Modern Orthodoxy. Feit also suggests that contemporary evolutionary theory should not pose a problem to anyone steeped in traditional Jewish sources. Troster, a Conservative rabbi who has written on science and Judaism and is active on Jewish environmental issues, argues that neo-Darwinism provides a serious challenge to traditional notions of divine action in the world. Drawing partially on theological resources outside Jewish tradition he seeks to develop a new direction for Jewish reflections about divine action in an evolving world. In these papers Feit and Troster suggest alternative approaches for those grappling with problems at the interface between evolution and Judaism.

As intellectual problem setting is a personal matter, it would be inappropriate for the editors of this volume to specify how readers should reflect on the contemporary challenges raised by the scientific theory of evolution. However, it is clear from this collection of essays that Judaism offers a range of theological and philosophical resources addressing such crucial issues as the nature of creation and the relationship between God and humankind. Some will also look outside Jewish tradition—to other theological and philosophical writings and, perhaps, to such areas as feminism or environmentalism. The only caveat we must enter is that the enquirer should be well-informed about both evolution and Judaism.

7

Crisis Management via Biblical Interpretation: Fundamentalism, Modern Orthodoxy, and Genesis

Shai Cherry

American Judaism is suffering a crisis of faith. Since the nineteenth century significant numbers of Jews have drifted away from traditional religion. By 1990, approximately 20 percent of people who identified themselves as having been born Jewish were no longer prepared to identify themselves as being of the Jewish religion.[1]

This tide of disbelief that has swept over educated American Jews has recently motivated three Modern Orthodox physicists to try to protect and promote Orthodox claims of the divinity of the Torah by advancing ostensibly literal readings of the first chapter of Genesis.[2] Of course, as the ancient rabbis well knew, and as many contemporary literary critics insist, any single text can generate multiple interpretations.[3] Yet in propounding their inventive interpretations of the Torah, these Modern Orthodox physicists are primarily concerned with shoring up the traditional claim of the Torah's divine provenance—a claim that has been undermined since the rise of biblical criticism in the second half of the nineteenth century.

In order to understand this study in crisis management, it is helpful to begin by looking outside the Jewish community. Nancy T. Ammerman has argued that modern Protestant fundamentalists in America are innovative in claiming that the Bible is inerrant at a time when there is no longer a

Gerald Schroeder warmly supported my initial work on this project. Our disagreements are for the sake of heaven.

1. *Highlights of the CJS 1990 National Population Survey* (New York: Council of Jewish Federations, 1991), 6.
2. I refer to this hermeneutic as *ostensibly literal* to emphasize both the intention of our three Modern Orthodox physicists and the rejection of their conclusions by contemporary biblical scholars who similarly seek plain-sense readings of scripture.
3. See, for example, Sanhedrin 34a; David Stern, *Midrash and Theory: Ancient Jewish Exegesis and Contemporary Literary Studies* (Evanston, IL: Northwestern University Press, 1996), 15–38.

consensus that the literal meaning of the Bible is true.[4] In rabbinic Judaism, by contrast, literal or plain-sense readings of Scripture never enjoyed a privileged status, especially regarding the account of Creation.[5] As Charles Liebman has pointed out, "Judaism . . . also believe[s] in the inerrancy of sacred scripture. But the internal debate centers on the meaning or interpretation of scripture."[6] Thus, the ostensible literalism of our Modern Orthodox physicists is doubly innovative.

Concern with the plain sense of Scripture became a communal issue in the Middle Ages, as rabbinic Jews struggled with Karaites, Christians, and Muslims in order to determine the straightforward, contextual meaning of the Bible's narratives and laws. For example, in this polemical context, Jews insisted that Isaiah's suffering servant was not a prophecy pointing toward the crucifixion of Jesus hundreds of years after the text was composed, but a metaphor for Israel within the political milieu of the ancient world and explicitly identified as the servant within the text itself (Isaiah 41:8). Moreover, in the last two hundred years there has been renewed attention to determining the plain sense of Scripture. One reason for this shift was the haskalah (Jewish enlightenment). Under the influence of the haskalah, the traditional emphasis on Talmudic *pilpul*, or casuistic reasoning, was largely replaced by a preference for the simple, contextual meaning of traditional texts.[7] Additionally, the haskalah emphasized the value of studying the Bible, as opposed to the Talmud and its interpretations.[8] Finally, beyond the traditional Jewish world, the academic study of the Hebrew Bible, which purports to concern itself exclusively with the Bible's meaning in its ancient Near Eastern context, was taking root. Thus, since the second half of the nineteenth century—that is, roughly since the publication of Charles Darwin's *On the Origin of Species* (1859)—Jewish thinkers have tended to pay

4. Nancy T. Ammerman, "North American Protestant Fundamentalism," in *Fundamentalisms Observed*, ed. Martin E. Marty and R. Scott Appleby (Chicago: University of Chicago Press, 1991), 14–16.

5. The aversion to an exclusively plain-sense reading of Jewish scriptures is predicated on the rabbinic assumption that the Torah is cryptic. See James Kugel, *The Bible as It Was* (Cambridge, MA: Belknap Press, 1997), 18–19. On Creation, in particular, see Mishnah Ḥagigah 2:1.

6. Charles S. Liebman, *Deceptive Images: Toward a Redefinition of American Judaism* (New Brunswick: Transaction Books, 1988), 44.

7. Emmanuel Etkes, "Immanent Factors and External Influences of the Haskalah Movement in Russia," in *Toward Modernity: The European Jewish Model*, ed. Jacob Katz (New Brunswick: Transaction Books, 1987), 14–15.

8. Ibid. This trend began in the 1780s with Moses Mendelssohn's translation of the Bible into German.

increasing attention to the simple sense of the Bible. This confluence sometimes translated into a synthesis of science and Torah, in which certain aspects of natural history were read into Genesis.[9] In the twentieth century, however, such syntheses were the exception rather than the rule.[10]

However, in the early 1990s three books were published offering scientific and ostensibly literal readings of the opening chapter of Genesis. None of the authors of these works was formally trained in the history of Jewish thought. Each was educated in the United States in physics—not biology!—and each identified with that hybrid class of contemporary Jewry, Modern Orthodoxy.

> [Modern Orthodoxy] tried to follow in a temperate way the path of culture contact embarked upon by the maskilim [adherents of the haskalah], valuing reason, science, and secular education; but by maintaining strong ritual and communal links to the traditional Jewish community, law and faith, they hoped to resist the assimilationist forces which had swept away their forebears. . . . Their adapted acculturation has inclined them to synthesize or compartmentalize their lives.[11]

While twentieth-century Jewish theologians have tended to compartmentalize science and the Torah,[12] our Modern Orthodox physicists synthesized them. Although there are historical precedents for attempting to synthesize Genesis with natural history, most notably by the towering figure of Jewish medieval philosophy, Moses Maimonides, proponents of this strategy did not claim to rely on literal readings of Genesis.[13] Instead, traditional Jewish thinkers regularly invoked rabbinic and kabbalistic views in discussing the Creation. By contrast, when our Modern Orthodox physicists sparingly invoke postbiblical sources, they claim that these later traditions

9. For examples of this phenomenon, see Michael Shai Cherry, "Creation, Evolution, and Jewish Thought," Ph.D. dissertation (Brandeis University, 2001), ch. 2.

10. Modern commentaries that focus on the contextual meaning, but without dealing with natural history, include David Zvi Hoffmann, *Book of Genesis* (in Hebrew) (Bnei Brak: Nezach, 1969), esp. 11; Umberto Cassuto, *A Commentary on the Book of Genesis* (Jerusalem: Magnes Press, 1944). See also Mordechai Breuer, "Faith and Science in Biblical Interpretation" (in Hebrew), *De'ot* 11 (1961): 18–25. For a similar approach, see Joseph B. Soloveitchik, *Lonely Man of Faith* (New York: Doubleday, 1992), 7.

11. Samuel C. Heilman and Steven M. Cohen, *Cosmopolitans and Parochials: Modern Orthodox Jews in America* (Chicago: University of Chicago Press, 1989), 17–18.

12. See Shai Cherry, "Three Twentieth-Century Jewish Responses to Evolutionary Theory," *Aleph: Historical Studies in Science and Judaism* 3 (2003): 247–90.

13. Moses Maimonides, *The Guide of the Perplexed*, 2 vols., trans. Shlomo Pines (Chicago: University of Chicago Press, 1963), 2:30, 351. Maimonides, in a frequently cited rejection of literalism, proclaimed, "Nor are the gates of figurative interpretation shut in our faces." 2:25, 327.

reveal the true meaning of the literal words of Genesis. This claim is consonant with the rabbinic belief that an Oral Torah was given to Moses at Mount Sinai, simultaneous with the giving of the Written Torah, the five books of Moses.[14] This Oral Torah was then transmitted through the generations until it was committed to writing at different times in Jewish history.

The ostensibly literal hermeneutic of our Modern Orthodox physicists echoes the rhetoric of North American Protestant fundamentalists who tend to rely on the King James translation and do not appeal to traditions of biblical commentary. Given the similarity between the rhetoric of Protestant fundamentalists and that deployed by our Modern Orthodox physicists to defend and promote their faith in the existence of God and the divinity of the Torah, it is appropriate to apply the term "fundamentalist" to both groups.[15] Furthermore, their selective application of this doubly innovative, ostensibly literal hermeneutic to aggadic (nonlegal) passages of Torah parallels a strategy used by ultra-Orthodox Jews. Thus, although our three authors would doubtless bristle at being labeled fundamentalists, their hermeneutical strategies are strikingly similar to those deployed by both Protestant and ultra-Orthodox fundamentalists.[16]

Nathan Aviezer and Genesis

Nathan Aviezer received his Ph.D. in physics from the University of Chicago and, in 1967, immigrated to Israel. He currently holds a chair in the department of physics at Bar-Ilan University, where he is a colleague of the physicist Cyril Domb, who coedited *Challenge: Torah Views on Science and Its Problems*[17] and also wrote the introduction to Aviezer's *In the Beginning . . . : Biblical Creation and Science.*[18] Domb clearly concurs with Aviezer's view that modern science reveals and confirms the literal meaning of Torah.

14. Sifra, Be-Ḥukkotai 2:8.

15. In his comparative analysis of fundamentalism, Bruce B. Lawrence suggests that one trait common to fundamentalists is that they "claim to derive authority from a direct, unmediated appeal to scripture." *Defenders of God: The Fundamentalist Revolt against the Modern Age* (San Francisco: Harper and Row, 1989), 100.

16. A discussion of the problematics of the term "Jewish fundamentalism" can be found in Michael Rosenak, "Jewish Fundamentalism in Israeli Education," in *Interaction of Scientific and Jewish Cultures in Modern Times*, ed. Yakov Rabkin and Ira Robinson (Lewiston, NY: Mellen, 1995), 374–75.

17. Aryeh Carmell and Cyril Domb, eds., *Challenge: Torah Views on Science and Its Problems*, 2nd ed. (Jerusalem: Association of Orthodox Jewish Scientists/Feldheim, 1978).

18. Nathan Aviezer, *In the Beginning . . . : Biblical Creation and Science* (Hoboken, NJ: Ktav, 1990).

Aviezer's book is structured as a biblical commentary that presents the data of natural history "day" by "day." Although he stresses the literal meaning of the Torah, Aviezer prefaces his analysis with the admission that a biblical day must be interpreted figuratively as a phase or period of time.[19] Aviezer also acknowledges that this interpretation was propounded by the rabbis of the Talmud. However, Aviezer is here being inconsistent since throughout his writings he has repeatedly asserted his commitment to a literal reading of the Bible's Creation narrative. Aviezer is clearly aware that he is being innovative in presenting such a literal rendition of the science of Genesis, but he feels justified in doing so by appealing to those advances in modern science that enable us to understand Genesis better when interpreted literally.

Before Aviezer offers his own exposition of Genesis, he attempts to undermine the credibility of Darwinian evolution by citing a number of prominent scientists who have admitted the lack of transitional species in the fossil record or have emphasized punctuated equilibrium in preference to Darwin's insistence on gradualism. Others cited by Aviezer have adopted Alvarez's impact theory of mass extinctions, as opposed to natural selection.[20] Identical strategies have been employed by Protestant fundamentalists in their attempts to discredit Darwinism, their logic being that if Darwinism is wrong, the Bible must be right. As Raphael Falk has already demonstrated that Aviezer has fatally distorted the scientific arguments, I will turn instead to a critique of Aviezer's biblical commentary.[21]

Aviezer begins by differentiating the two verbs used in Genesis to describe divine activity. "Creation implies the formation of something fundamentally new, either physically (as in creation *ex nihilo*) or conceptually (as in the creation of a new kind of life). By contrast, the process of making implies the fashioning of something complex from something simple."[22] He then leapfrogs from philology to biology: "This interpretation of the biblical text, which does not deviate from the literal meaning, incorporates the notion that present-day animals developed from earlier species."[23]

There are, he points out, three instances of the verb *to create* (*bara*) in the first chapter of Genesis. The first, in verse one, describes Creation ex nihilo

19. Ibid., 1.

20. Ibid., 54–55, 59–60. See also Nathan Aviezer, "Misreading the Fossils: The Dark Side of Evolutionary Biology," *Bekol Derakhekha Daehu* (hereafter, *BDD*) 2 (1996): 19–30, reprinted in Nathan Aviezer, *Fossils and Faith* (Hoboken, NJ: Ktav, 2001), ch. 14. Chapters 16–18 of *Fossils and Faith* also attempt to discredit Darwinism.

21. Raphael Falk, "In the Beginning?" (in Hebrew), *Alpayim* 9 (1994): 133–42.

22. Aviezer, *In the Beginning*, 60.

23. Ibid.; see also 71, 101.

of all matter in the universe. The second, in verse twenty-one, is interpreted by Aviezer as the ensoulment of matter and the creation of animal life, beginning with the large sea creatures that he identifies as Ediacaran fauna. The third instance of *to create* relates to modern humans being graced with the divine image. According to Aviezer, the speciation that occurred between the second and third of these creations involved the "fashioning [of] something complex from something simple." Aviezer's version of theistic evolution posits a divinely guided transmutation of species over deep time. According to Aviezer, Genesis literally describes exactly that.

According to Aviezer, the swarming waters with which animal life began (Genesis 1:20) refer to the explosion of life forms constituting the "dramatic events that occurred at the beginning of the Cambrian period." The winged creatures created on day 5 included not only birds, which science tells us evolved from land animals (i.e., products of day 6), but also winged insects. Aviezer explains that the geological division between days 5 and 6 represents the cusp between the Paleozoic and Mesozoic periods. He attributes the relative fecundity of marine and insect life (day 5) as opposed to terrestrial life (day 6) to the divine blessing issued to the creatures on day 5 to "be fruitful and multiply" (Genesis 1:22). The nonhuman creatures of day 6 received no such blessing.[24]

As for the evolution of hominids, Aviezer again cites the views of prominent scientists to conclude that "*Homo erectus* did not undergo *any* significant evolutionary changes throughout the million years of his existence."[25] After presenting a series of scientists' statements that reveal their perplexity about the sudden emergence of modern man with revolutionary new capabilities, including artistic representation, tool making, and ritual activity, Aviezer presents his solution:[26]

> It is particularly meaningful that Modern Man is intellectually and culturally *so vastly superior* to his closest relative, the extinct Neanderthal Man, even though both species are *physically very similar*. They possess the same average brain size as well as the "same postural abilities, manual dexterity, and range of movement." Scientists have no explanation for the enormous difference between Modern Man and Neanderthal Man. They are puzzled by the fact that a comparison between the physical features of the two species "does not suggest any differences in intellectual or behavioral capacities." In contrast to the secular scientist, one who believes in God *does* have an explanation for the

24. Ibid., 84–86.
25. Ibid., 92.
26. Ibid., 92–99

> uniqueness of Modern Man—an explanation that is written in verse 1:27: "And God created Man in His image."[27]

This human, now endowed with the divine image, was then blessed by God and told to "fill the earth and master it" (1:28). Aviezer identifies this blessing with the agricultural or Neolithic revolution.[28] This blessing brought with it the origins of agriculture, animal husbandry, metalworking, the wheel, language, ceramic pottery, weaving, music, advanced architecture, and more.

In his latest book, *Fossils and Faith*, Aviezer recognizes a discrepancy in his earlier chronology. He had previously identified the human of Genesis 1 with the Neolithic period, which occurred some 10,000 years ago. But this chronology is clearly incompatible with Jewish tradition, which maintains that only about 6,000 years have elapsed since the creation of Adam, leaving a 4,000-year hiatus. Aviezer reconciles the dates by arguing that the humans of Genesis 1 refer to all of humanity receiving the divine image, whereas the Adam of the second chapter of Genesis was placed in the Garden of Eden only 6,000 years ago. The Jewish calendar, therefore, began only with the Adam of the second chapter of Genesis.[29]

Aviezer allows for the transmutation of species, divinely guided, over deep time. In that regard, he is no young-earth special creationist. His agenda is to convince his Modern Orthodox readers that there are no contradictions between demonstrated science and his scientifically inspired, ostensibly literal reading of Genesis. Thus, shoehorned into nine verses of Genesis are to be found the Cambrian explosion, the Paleozoic and Mesozoic periods, the disappearance of the Neanderthal man, the rise of modern man, and the Neolithic revolution. Through modern science, Aviezer insists, "the biblical text *can* be understood in its literal sense."[30] Indeed, without modern science, the biblical text would remain opaque. "Modern science has become a significant element in the strengthening of ancient faith."[31] Aviezer wants to strengthen that ancient faith in God and the divine origin of Torah in order to bolster the commitment of Modern Orthodox Jews.

27. Ibid., 100–101. The claims in this passage are substantiated in three footnotes. Notice that Aviezer conflates belief in God with the belief in his scientific reading of Genesis 1:27.

28. Ibid., 101.

29. Aviezer, *Fossils and Faith*, 49.

30. Aviezer, *In the Beginning*, 2.

31. Aviezer, *Fossils and Faith*, 8.

Gerald Schroeder and Genesis

Gerald Schroeder received his doctorate in earth sciences and nuclear physics from the Massachusetts Institute of Technology. Immigrating to Israel in 1971, he lectured for Aish HaTorah, an ultra-Orthodox yeshiva in Jerusalem. Although Schroeder lectures to secular Jews on behalf of an ultra-Orthodox yeshiva, he writes for a general audience. Aviezer's books are published by Ktav Publishing, a Jewish house, while Schroeder uses secular publishers. Although both focus on Genesis and refer to Jewish Bible commentators, Schroeder cites several New Testament verses and frequently mentions the Judeo-Christian tradition.[32] In fact, Schroeder reached out specifically to a conservative Christian audience in a series of television recordings that were later transcribed.[33] Schroeder's texts also differ from Aviezer's in that Schroeder is not concerned with parochial Jewish observance, but rather with universal ethical behavior.[34]

Yet, despite these differences, their methodologies are remarkably similar: they both analyze Genesis using an ostensibly literal hermeneutic. Ironically, in an effort to dissociate himself from American fundamentalism and its rejection of both evolution and deep time, Schroeder argues against the "sophistry of biblical literalism."[35] Schroeder, indeed, relies on later Jewish biblical interpretations, specifically those of Maimonides and Naḥmanides. But for Schroeder, these interpretations merely articulate what the Bible tersely, but literally, states.

Schroeder, like Aviezer, believes that Genesis contains nothing inimical to science. On the contrary, modern scientific knowledge allows us to understand the true, literal meaning of the Creation narrative. Before addressing the specific details of evolution, Schroeder, again like Aviezer, must explain the seeming discrepancy between the six days of Creation, as recorded in Genesis, and the fifteen billion years that cosmologists calculate as having elapsed since the Big Bang. Schroeder, however, disparages figurative

32. Gerald L. Schroeder, *Genesis and the Big Bang: The Discovery of Harmony between Modern Science and the Bible* (New York: Bantam, 1990) [hereafter, *GBB*], 18, 26, 57, 63; Schroeder, *The Science of God: The Convergence of Scientific and Biblical Wisdom* (New York: Free Press, 1997) [hereafter, *SoG*], 82, 178; Schroeder, *The Hidden Face of God: Science Reveals the Ultimate Truth* (New York: Simon and Schuster, 2002) [hereafter, *HFG*], 21–22.

33. *Genesis One: Dr. Gerald Schroeder with Zola Levitt* (Dallas: privately published, 1998).

34. *GBB*, 8 f., *SoG*, 3, 201 f. Aviezer, by contrast, writes that "one merits God's blessings by observing His commandments," *Fossils and Faith*, 124.

35. *SoG*, 12. See also *GBB*, 16.

readings of the six days as an "easy explanation."[36] In his most important contribution to the synthesis or concordance of science and Torah, Schroeder explains how the universe was literally created in six days. Employing Einstein's theory of relativity and time dilation, Schroeder argues that from the perspective of the forward-rushing cosmos, six days is equivalent to fifteen billion years from the perspective of the backward-looking earth.[37] We do not need to enter into the details of Schroeder's argument or speculate how recent downward revisions of the universe's age would affect his calculations. What is crucial is that Schroeder presents a reconciliation between the apparently irreconcilable literal reading of Genesis and the findings of modern science.

In discussing the fifth and sixth days of Creation, Schroeder offers his own concordance. He agrees with Aviezer that the winged animals of day 5 are water insects.[38] But the "great sea monsters" of verse 21 are not Ediacaran fauna, measuring a measly forty-five centimeters. They are "the big reptiles" that Schroeder identifies as dinosaurs![39] In discussing the fossil record of dinosaurs dating from day 5, Schroeder claims that there exists only one single example of an animal that represents a transition between the basic levels of phylum or class—the archaeopteryx.[40] This creature, which Schroeder identifies as the *tinshemet* of Leviticus 11:18 and 11:30, lived 150 million years ago and is considered to be a transitional form between reptiles and birds. In the lists of animals that are not fit for Israelite consumption, the *tinshemet* appears under the categories of both birds and reptiles.

> In the entire Bible, there is the one reference to an animal that falls into two categories, the *tinshemet*. In the entire fossil record there is one fossil that falls exactly midway between two classes of animals, the archaeopteryx. And both the archaeopteryx and the *tinshemet* are part reptile, part bird. It is the "link" that never was missing.[41]

Schroeder does not address the issue of why a creature that had been extinct for 150 million years should appear in a list of foods forbidden to Israelites.

Moving to day 6, Schroeder posits that there were hominids, which lacked the image of God, present at the time of Adam's birth. The existence of these

36. *GBB*, 29. Nowhere in Schroeder's books does he specifically mention Aviezer.
37. Ibid., 34–55 and *SoG*, 41–71.
38. *SoG*, 94.
39. Ibid., 193.
40. Ibid., 95.
41. Ibid., 96.

prehistoric men, Schroeder claims, is attested by Maimonides in his *Guide of the Perplexed*.[42] Yet Schroeder's literal hermeneutic has led him astray. Maimonides does indeed refer to men without human form and calls them animals. However, Maimonides was not discussing zoology. For Maimonides, form was an Aristotelian, not a physical, concept. The form of human beings is intellectual perfection.[43] Thus Maimonides maintained that among humans, very few have attained human form, i.e., intellectual perfection. Instead, many people were to be compared to animals.[44]

Schroeder hints that some Cro-Magnon individuals, whom he identifies as the giants of Genesis 6:4, were transformed into modern man through divine ensoulment. The indicator of the emergence of modern man, Schroeder tells us, is the advent of writing. "Archeologists date the first writing," according to Schroeder, "at five or six thousand years ago, the exact period that the Bible tells us the soul of Adam, the *neshama*, was created."[45] Does that mean that Adam could write? Schroeder cites Onkelos's ancient Aramaic translation of the Pentateuch, which, he claims, substantiates this claim. Genesis 2:7 describes the human becoming a *nefesh ḥayyah*, which the New Jewish Publication Society's *Tanakh: The Holy Scriptures* translates as "living being." Onkelos translated that term into Aramaic as *ruaḥ memalela*, which Schroeder then translates as "communicating spirit."[46] Communication takes place through speaking and writing. Onkelos's translation should, however, be more accurately rendered as "speaking spirit." But, in an effort to include the notion of writing in Onkelos's translation of the creation of humanity, and thus offer further evidence for the convergence of science and Torah, Schroeder interpreted Onkelos liberally rather than literally.

Schroeder understands that there are two major theological problems with claiming that God controls the path of natural history: providence and theodicy. He also understands that any attempt to rehabilitate the Torah as a voice of divine authority must account for the apparent evils in both natural and human history. Indeed, in his first book, Schroeder concedes that evolution does not trace a straight path from the Cambrian explosion to modern man. Schroeder compares the dead-ends and reversals of natural history to the journey of the ancient Israelites as recorded in the Bible. When the

42. *GBB*, 151, citing Maimonides, *Guide* 1:7.

43. Maimonides, *Guide* 1:1, 22–23.

44. Ibid., 1:7, 33.

45. *SoG*, 143. See also 130, 163.

46. Ibid., 145.

Israelites were escaping from the Egyptians, God had them double back on their path (Exodus 13:17, 14:2). "From the Bible's worldview," Schroeder writes, "there is no reason to expect that the path from the primitive amoral aquatic animals of day five, 530 million years ago, to sentient, moral humans near the close of day six, 6,000 years ago, would be any more direct than was the trek of the Israelites from Egypt to Sinai."[47] The Torah tells us how God operates in Jewish history and we can extrapolate from that information to an understanding of how God operates in natural history.

Schroeder recognizes that this does not solve the problem of teleology or divine providence, either in natural history or in human history. Nevertheless, by emphasizing the parallel between Jewish history and natural history he imposes his own perspective on the problem. Even if we accept that while divine control may appear enigmatic to the participants, it may nevertheless seem rational when considered from the divine perspective of the overall scheme. If we accept this dichotomy of perspectives we may be less troubled by what appear to us as dead ends.[48] This parallel is crucial for Schroeder's argument because he acknowledges that natural history has been far more chaotic than the harmony and order that is suggested by the first chapter of Genesis.

In Schroeder's second book, *The Science of God*, he moves from teleology and providence to theodicy. Unlike Aviezer, Schroeder does not ignore these issues, since he appreciates that any religious discussion of evolution must confront them. Schroeder equates the Lurianic concept of *tzimtzum* (divine withdrawal) with the biblical notion of the hiding of God's face.[49] This act of divine withdrawal has allowed room for nature to take its meandering course and also allows for human free will. Without entering into the theological nuances that exercised Jewish metaphysicians for centuries, Schroeder simply claims that the Jewish God is both transcendent and immanent. The Creator of natural laws periodically intervenes in the world when nature has gone awry.[50]

> Obviously, an omnipotent creator could remove all randomness from nature. Crippled children and hereditary diseases would be no more. But the price would be too high. Without some degree of randomness, all events and all choices in the universe would be totally predetermined by unyielding laws of nature, the physics and chemistry of all reactions. We would be mere robots. . . .

47. *SoG*, 79–80. See also *GBB*, 170–72.

48. *SoG*, 79. The argument is circular. We first need to assume that the Torah is divine in order to accept the analogy, which serves as evidence that the Torah is divine.

49. Ibid., 17, 171.

50. Ibid., 14–15 and 78–80.

> Freedom in nature—so that not every stellar system is a life-nurturing solar system—and freedom of will—so that a given stimulus produces a variety of responses—are traits of the divine contraction, the *tsimtsum*, which brought our universe into existence. Whether *tsimtsum* is divinely essential in universe formation or a deliberately chosen aspect of the design of our universe is a question we cannot answer. It is, however, the reality of our existence.[51]

In his latest book, *The Hidden Face of God*, Schroeder has taken the parallels between human history and natural history to yet another level. He reminds his readers that, according to the Torah, God destroyed the world in a flood. His conclusion is that "the universe is far from perfect. . . . What is important here is the biblical message being presented: that the Divine design of the world was flawed, required Divine retuning, and the Tuner acknowledged this need. According to the Bible, intelligent design, even at the level of the Divine, is not necessarily perfect design."[52] The Torah thus discloses a world where divine guidance is not always apparent, where the exercise of free will necessitates the possibility of evil, and where divine design is imperfect. Schroeder then develops these points in order to arrive at a Jewish theology of nature in which teleology is opaque, extinctions are legion, and design flawed.[53]

Once again we have an example of the scientist "revealing" the true meaning of the Torah.[54] Schroeder hopes that if he can shock his readers with his claim that the Torah recorded scientific facts thousands of years prior to their corroboration by science, then perhaps they will be more likely to obey the biblical commandment to love their neighbors.[55] By using a conceptual and historical hodgepodge of biblical, midrashic, Aristotelian, and kabbalistic sources, Schroeder opportunistically constructs a Creation narrative whose ostensibly literal interpretation roughly corresponds to the scientific account of the physical universe.

Judah Landa and Genesis

Like Aviezer and Schroeder, Judah Landa was trained in the United States as a physicist. He teaches high school physics in Brooklyn, where for many years

51. Ibid., 169–70. Eliezer Berkovits anticipated this argument in his *Faith after the Holocaust* (New York: Ktav, 1973), 106.

52. *HFG*, 10. Schroeder, unlike advocates of Intelligent Design, explicitly and unabashedly tethers his science to the biblical text.

53. *GBB*, 115, 136, 149; *HFG*, 11–12.

54. *SoG*, 130.

55. Ibid., 82, 201–2.

he was at the Yeshiva of Flatbush. Unlike Aviezer and Schroeder, however, Landa does not present a pious alternative to Darwinism. Instead, his *Torah and Science* is a curious combination of science and biblical interpretation designed not to persuade his readers that *Genesis* anticipated Big Bang cosmology and biological evolution, but that there is no necessary conflict between Genesis and science.[56] While our first two scientists struggled to discredit Darwinism and promote the anthropic principle, Landa does neither.

Landa lays out the proofs for deep time—a commitment shared by all three of our physicists.[57] He concludes that the estimates of the earth's age, about 4.6 billion years, and the universe's age, about 15 billion years, have been scientifically demonstrated so that no reasonable person can doubt them. In a surprising move, Landa then extrapolates from his own experience and declares that "most Orthodox Jews" insist on interpreting the biblical story of Creation literally such that "all was created in six days."[58] He also asserts that the rabbis of the Talmudic period insisted on a literal six-day Creation scheme.[59] Why does Landa repeatedly emphasize the consensus over interpreting the six days literally but avoid all mention of rabbinic precedents for reading the days of Creation figuratively?[60] In order to suggest an answer to this question, we must first review Landa's discussion of evolution.

As Darwin claimed, variation always occurs within any animal species. Since the modern synthesis, scientists have attributed many of these variations to genetic mutations. Mutations, though frequently described as random, are not acausal. Landa argues that it is only through our ignorance that evolution appears random—it is, in actuality, a deterministic process of cause and effect.

> First, evolution is anything but a random process. Every step of the way is supposed to be guided by the laws of nature, particularly those that relate to the behavior of atoms, molecules, and subatomic particles. Mutations appear random to us because we don't know enough about them to control and predict their occurrence and outcome. But every mutation has a cause and produces an effect, both of which are propelled by the laws of the universe. . . . And the process of selection between species was directed by the eminently reasonable principle of survival of the fittest.[61]

56. Judah Landa, *Torah and Science* (Hoboken, NJ: Ktav, 1991).
57. Ibid., 263–81.
58. Ibid., 290.
59. Ibid., 297, 313, 318, 325.
60. For a presentation of rabbinic statements, see Eli Munk, *The Seven Days of the Beginning* (Jerusalem: Feldheim, 1978).
61. Landa, *Torah and Science*, 293.

Thus far, most biologists would generally agree with Landa's description of evolution. Yet in the following paragraph, Landa begins to elide efficient causes with final causes.

> It is the complexity of the phenomena and our incomplete understanding of nature's ways . . . that give evolution the appearance of randomness. The reality of the situation, however, is that nature continues to do "its thing" whether or not we fully understand its ways or can keep track of the details. Evolution was designed and guided, just as the putting together of words and sentences into book form is accomplished only by design and guidance. A book is designed by its author; evolution was (and continues to be) designed by the laws of nature (which, in turn, were designed, we believe, by God).[62]

In this depiction of God indirectly guiding evolution through natural law Landa conflates the notions of efficient causality and final causality. This conflation undermines Darwin's revolutionary insight, since Darwinian evolution postulates antecedent causes, but not prospective purposes. Evolution is not teleological. Yet the analogy that Landa employs between God and the author of a book is unacceptable to orthodox Darwinians since an author designs a book with foreknowledge of the development of both characters and plots. Even if the climax of the plot is unknown to the author at the commencement of the project, she still controls the environment and characters throughout the process in order to achieve her desired end. Darwinism precludes just this kind of teleological overview.

Yet Landa is not consistent, and at some points he sounds like an orthodox Darwinian: "The life forms found on earth today constitute the particular ink-blot pattern that emerged from the original atmospheric soup mix (the cosmic ink spill)."[63] Ink being spilled is a process without design. Landa also agrees with modern science that "it is more reasonable to compare life on earth, with all its plusses and minuses, to an irregular pattern of ink than to a book."[64] He argues the point convincingly by appealing to those flaws in nature that are difficult to reconcile with a benevolent and omnipotent deity.[65] If so, how can the ink blot be read as a book?

> Finally, the two theories, evolution and design by God, are not mutually exclusive. Science makes no comment, and claims no competence to comment, on the origin of the laws of nature or on the origin of the matter and

62. Ibid., 293–94.
63. Ibid., 294.
64. Ibid.
65. Ibid., 295–96.

> energy of the universe. God may very well have designed the laws of the universe and the earliest forms of matter and energy with particular life-forms as end-products in mind. Evolution and natural selection may be the vehicles he chose and designed to achieve His purposes. *The real issue,* thus, is not whether life as we know it emerged by design and purpose or by accident, without planning. Instead the issue is: Did life evolve over the course of billions of years, guided by laws of nature and natural selection, or did it appear suddenly, 6,000 years ago, with natural selection playing no role?[66]

Landa is tentative when he writes that God may have had a particular telos in mind when initially devising the laws of nature. He simply makes the incontrovertible claim that science cannot prove the nonexistence of God or of final causes. Even scientific materialists would agree with that. Modern science is the search for efficient causes, but final causes are to be left to metaphysicians and theologians.[67]

Landa accepts the basic ideas of Darwinian evolution. He does not argue for final causes; he simply asserts that they cannot be ruled out, even though natural selection is in tension with traditional notions of teleology. His references to apparent flaws in nature are presented as arguments against final causes. Like Aviezer, he nowhere attempts to offer a theodicy or deal with the problem of providence. Instead, Landa asserts, as a matter of faith, that God is the author of the laws of nature and of evolution. He also states that science cannot adjudicate Judaism's claims for final causes or providence.

Divine providence and teleology are not objects of scientific inquiry; therefore, they are not the "real issue" for Landa. By contrast, Aviezer and Schroeder are preoccupied with precisely what Landa has dismissed, namely, whether or not life evolved teleologically. When Landa uses the term "real issue" in the above passage, he means the pressing problem of negotiating between religious claims and scientific facts. The discovery of final causes cannot be a real issue, because science does not address final causes. Once Landa has disposed of the knotty problem of teleology by invoking the methodological limitations of science, he can focus on the relatively simple issue of interpreting the six days of Creation. From Landa's perspective, the Creation narrative is the real issue because it addresses claims made by science.

By (incorrectly) claiming that a consensus existed among the Talmudic rabbis that "days" should be interpreted literally, he shifts attention away from the complicated problems of theodicy and teleology to the far less complicated problem of biblical interpretation. In other words, Landa relocates

66. Ibid., 296. Italics added.
67. Ibid., 345.

the problems of evolution from the theological and scientific to the literary and hermeneutic. To resolve this literary issue, Landa offers the following interpretation of the first chapter of Genesis. With corroboration from such traditional commentators as Rabbis Shlomo ben Yitzḥak (Rashi) and Abraham ibn Ezra, Landa argues that

1) since the first word of Genesis is grammatically in the construct state, only light was created on the first day,
2) *bara* does *not* mean creation out of nothing,
3) there were elements, such as the earth and water, that were in existence prior to day one, and
4) the Torah does not purport to tell us the sequence of Creation.[68]

With those four claims, based on a literalist hermeneutic, Landa writes a scientific commentary on Genesis. Since there existed primordial matter prior to day 1, time must have existed in some form prior to the Creation described in the opening verses of Genesis. Landa buttresses his literal reading of Genesis by drawing on rabbinic commentary that supports the notion of orders of time prior to the beginning of the cosmos in Genesis 1:1.[69]

Landa next argues that the idea of twenty-four-hour days in Genesis is nonsensical.

> But if light alone was created on the first day, as Ibn Ezra says and Rashi's translation makes reasonable, the first day could not possibly have consisted of twenty-four hours. For the darkness that preceded the light of that day existed before anything was created in the biblical story. Well then, what delineates the beginning of that night? Nothing! That night stretches all the way back in time for God knows how long, with no apparent beginning. Only its end is defined—by the creation of light that started the morning of the first day. Since that night is part of the one day described in Genesis, the first day cannot be a twenty-four-hour interval.
>
> And why should any of the first three sets of night and day be construed as twenty-four-hour periods when the sun, moon, and stars were not created until the fourth day? . . .
>
> And if the first, second, and third cycles of light and darkness do not correspond to twenty-four-hour intervals, perhaps all six cycles are not meant to do so?[70]

Landa offers additional reasons, both logical and hermeneutical, why the days of Creation should be understood as other than twenty-four-hour

68. Ibid., 314–16. This is a summary of Landa's argument.

69. Ibid., 317.

70. Ibid., 319.

periods. Readers are reminded that in the Bible the plague of frogs in Egypt is described using the singular word "frog" (Exodus 8:2). By analogy Landa suggests that the use of the singular "day" in the biblical account of Creation implies not one day, but many days. His conclusion is that the biblical days are to be interpreted as eras, and that if the medieval rabbis who composed the traditional commentaries on the Torah were alive today, given the scientific proof of deep time, they would endorse such an interpretation.[71] What distinguishes Landa's argument from our other two Modern Orthodox physicists is that he appeals to the Torah to justify his reading of days as eras.

"Now that we have as much time as we need, before and during the six days of creation, for evolution to have taken its course, the heart of the conflict between Torah and science (creation vs. evolution) dissolves into nothingness," writes Landa. He continues: "All that remains to be done is to explore the rest of the story to see *how it could be made to fit into the evolutionary scheme of things*."[72] Landa then offers a facile concordance between science and Torah but pays little attention to the sequence of events. He justifies this maneuver by citing Rashi's comment that Genesis does not describe the order of Creation and by appealing to the traditional rabbinic principle that there is no chronological sequence in the Torah.[73] Landa, however, does not explain the methodology by which he accepts the rabbinic view on chronology but rejects their (supposed) consensus on a six-day Creation.

Unlike the other concordantists, Landa is not looking for prescience in the Creation narrative but rather for interpretations that neutralize the potential conflicts that science poses to a literal interpretation of Torah. Landa emphasizes that there is no conflict between the Torah and science, only between rabbinic interpretations of Torah and science. When the Bible is properly interpreted—that is, when his ostensibly literal hermeneutic is employed—the problems "dissolve into nothingness."[74]

Underpinning Landa's arguments is his concern with the future of Orthodox Judaism. If Jews accept that Judaism promotes the antiscientific claim that Creation occurred over six twenty-four-hour days, "those who are wavering in their commitment to the practice of Judaism, for one reason or another, or having given up that commitment, are considering returning to the fold," will be justifiably alienated.[75] Landa fervently hopes that his scientific interpretation of the Creation narrative "will invigorate and strengthen

71. Ibid., 318, 326.

72. Ibid., 321. Italics added.

73. Ibid., 321–23. See Pesaḥim 6b and parallels for the rabbinic maxim of asynchronism.

74. Landa, *Torah and Science*, 325, 347, 321.

75. Ibid., 326.

the practice of Orthodox Judaism and remove the dark cloud which has been hovering, needlessly, over the faith for decades."[76] If the Orthodox community, to whom his book is addressed, distances itself from a commitment to a young earth and instead embraces the scientifically demonstrated transmutation of species over deep time, two important changes will occur. First, those presently committed to Orthodoxy will gain a better understanding of science. Second, those not fully committed to Orthodoxy will have one fewer obstacle to overcome on their journey back "to the fold."

Conclusion

Our concordantists base their authority on their status as scientists, not theologians. It is the prestige that modern culture has invested in the scientific enterprise that allows these nontheologians to serve as interpreters of reality and of Genesis. To the extent that there has been an apotheosis of science in our secular world, scientists are our new priests. The secularism of modernity may help to explain why some religious scientists defend traditional theology, but it does not explain why these American Modern Orthodox scientists resort to an ostensibly literal hermeneutic in order to integrate science with Torah. Yet it is not this hermeneutic alone that makes our authors fundamentalists.

Our Modern Orthodox physicists qualify as fundamentalists because their literalism is selective. They do not, for example, apply literalist hermeneutics to the legal passages of the Torah. A Reuben sandwich that combines milk and meat is still off-limits, even though it does not involve seething a kid in its mother's milk (Exodus 23:19). This concern with aggadic material is also characteristic of the fundamentalism of the ultra-Orthodox, who, ironically, reject scientific education. As Michael Silber writes:

> By and large, the ultra-Orthodox ignored the mechanism which had guided Jews over the centuries in sifting, weighing, discarding and reconciling the multiplicity of *aggadic* statements that were often sharply at odds with one another. If in fact there is reason to designate ultra-Orthodox Judaism "fundamentalist," it is precisely because of its tendency to ignore the "tradition" of these traditions in favor of a literal reading. Thus, any one strand of tradition could always be seized upon and cited, no matter how extreme or marginal, because it did after all appear in the written sources.[77]

76. Ibid., 327.

77. Michael K. Silber, "The Emergence of Ultra-Orthodoxy: The Invention of a Tradition," in *The Uses of Tradition: Jewish Continuity in the Modern Era*, ed. Jack Wertheimer (New York: Jewish Theological Seminary of America, 1992), 61.

Like the ultra-Orthodox, who excavate isolated aggadic statements that lack the communal consensus of halakhah (Jewish law) and read those aggadot literally, our authors read the first chapter of Genesis literally, while ignoring other biblical accounts of Creation as well as the rich history of Jewish interpretation of Creation.

Is the influence of American Protestant fundamentalism or ultra-Orthodoxy sufficient to explain why our authors have adopted an ostensibly literal reading of Genesis?[78] Or, at a deeper causal level, are these Modern Orthodox physicists in America motivated by the same factors as Christian, or indeed other, fundamentalists? Two authorities on fundamentalism write:

> Religious fundamentalism . . . manifests itself as a strategy, or set of strategies, by which beleaguered believers attempt to preserve their distinctive identity as a people or group. Feeling this identity to be at risk in the contemporary era, these believers fortify it by a selective retrieval of doctrines, beliefs, and practices from a sacred past. . . . Moreover, fundamentalists present the retrieved fundamentals alongside unprecedented claims and doctrinal innovations.[79]

In the decades since the Shoah, with intermarriage rates climbing to 50 percent, many Jews—especially more traditional Jews—feel beleaguered. After World War I, some elements within the Protestant community experienced a similar sense that their civilization was crumbling. It was then that Protestant fundamentalism, itself an early twentieth-century phenomenon, linked special Creation to a literal hermeneutic of Genesis.[80] The Jewish version of creationism is more scientifically sophisticated than its Protestant counterpart, but fundamentalists in both religions share a sense of crisis about the future.

As important as the method of selective literalism is in establishing a connection between our Modern Orthodox physicists and both Protestant fundamentalists and ultra-Orthodox Jews, we must also consider their motivation. As Bruce Lawrence writes, "Fundamentalists were left as the last defenders of the fortress of faith. They were not defending simply themselves, or the church (synagogue, mosque) or scripture (Bible, Torah, Qu'ran). They were defending

78. On the influence of ultra-Orthodoxy on Modern Orthodoxy, see Charles Selengut, "By Torah Alone: Yeshiva Fundamentalism in Jewish Life," in *Accounting for Fundamentalisms: The Dynamic Character of Movements,* ed. Martin E. Marty and R. Scott Appleby (Chicago: University of Chicago Press, 1994), 236–63, esp. 259.

79. Martin E. Marty and R. Scott Appleby, "Introduction: A Sacred Cosmos, Scandalous Code, Defiant Society," in *Fundamentalisms and Society,* ed. Martin E. Marty and R. Scott Appleby (Chicago: University of Chicago Press, 1993), 3.

80. See Ronald L. Numbers, "The Creationists," in *God and Nature: Historical Essays on the Encounter between Christianity and Science,* ed. David C. Lindberg and Ronald L. Numbers (Berkeley: University of California Press, 1986), 391–423.

GOD."[81] Likewise, our authors defend God and the divinity of the Torah. Aviezer and Schroeder promote the anthropic principle, which they interpret as evidence of an intelligent Designer preparing the universe for human habitation, and shoehorn the facts of modern science into an ancient text, thereby arguing for its prescience and divine authority.

In attempting to restore faith in Torah, Aviezer and Schroeder are trying to reverse the general trend over the last three centuries of dismissing the Bible as an indubitable source of scientific knowledge. American Jews, including Modern Orthodox Jews, are much more likely to accept evolutionary theory than are other Americans. According to a 1982 Gallup poll, 44 percent of all Americans believe the world was created more or less as it currently is within the last ten thousand years.[82] Over 80 percent supported the inclusion of creationist theories in public school science curricula.[83] At about the same time as this Gallup poll, Heilman and Cohen conducted a survey of American Jews, most of whom identified themselves as Modern Orthodox. Among the Modern Orthodox in this sample, 85 percent agreed that their children should be taught the theory of evolution in school.[84] The figures were even higher for the nominally Orthodox and non-Orthodox Jews. Although these two polls were asking different questions, they suggest that a far higher percentage of American Jews accept evolution than do American non-Jews. Moreover, according to the 1990 National Jewish Population Survey, less than a quarter of those who identified as Jewish agreed that "the Torah is the actual word of God."[85]

We need to understand the motivation of Aviezer, Schroeder, and Landa against this background. They were targeting an American Jewish community that privileges science over the Torah as a source of scientific information. If Genesis could be shown to have anticipated Darwin or Einstein, then the Bible would regain an aura of truth that it had been losing since the advent of biblical criticism and modern science. If these Orthodox scientists could convince their readers that the Bible contains true knowledge of divine origin, then their goal of protecting Orthodox Judaism and promoting Orthodox halakhah will be nearer to fulfillment. Or so they hope.

Notwithstanding the commercial success of Aviezer's and Schroeder's books, the statistics for American Jewry evidence a further slide toward

81. Lawrence, *Defenders of God*, 107.

82. Ammerman, "North American Protestant Fundamentalism," 2. A 1997 Gallup poll arrived at the same figure.

83. George Gallup, "Public Evenly Divided between Evolutionists, Creationists," Los Angeles Times Syndicate, 1982 (press release). By 2000, that figure was down to 68 percent.

84. Heilman and Cohen, *Cosmopolitans and Parochials*, 158–59.

85. *Highlights of the CJS 1990 National Population Survey*, 30.

secularism.[86] While in 1990, 20 percent of those who identified themselves as having Jewish parents denied that they had any religion, by 2001 that figure had jumped to 32 percent.[87] Also pertinent is that 17 percent of the core Jewish population of the United States (those of Jewish parentage and those who have converted to Judaism) in 2001 considered themselves to be atheists or agnostics; for American non-Jewish adults, the figure was only about 1 percent. Moreover, a whopping 44 percent of those who identified their religion as Jewish described their outlook as secular or somewhat secular, compared to 16 percent among non-Jewish adults.[88]

Although it is impossible to determine the influence our authors have exerted on the Modern Orthodox community, the statistics suggest that Orthodoxy survived the 1990s relatively well. In fact, among the eighteen-to-thirty-four age cohort who were affiliated with a Jewish religious institution, more than one-third belong to Orthodox synagogues![89] Although the majority of Jews who were raised in Orthodox homes no longer identified themselves as Orthodox, there was close to a 50 percent increase within the Jewish population of those who did identify themselves as Orthodox: a rise from 7 percent in 1990 to 10 percent in 2000.[90] Despite the secularizing trend among American Jews, these statistics also suggest a countervailing tendency. The crisis has been mitigated, if not quite managed.

Aviezer wrote his books as a bulwark against the weakening of Modern Orthodoxy. Schroeder's more ambitious goals included outreach to Jews and Christians disaffected from the biblical tradition, whereas Landa hoped to persuade those in his own Orthodox world that his scientific reading of Genesis would not only increase scientific literacy among the Orthodox but also remove an obstacle for Jews whose commitment to Orthodoxy was only partial. Aviezer and Schroeder desired to turn an enemy into an ally with their claim that Genesis anticipates the findings of modern science; while Landa was more modest in trying to neutralize the threat of science by removing the apparent contradictions between science and Genesis (when properly interpreted). Yet all three writers engaged in three innovative practices that identify them as fundamentalists. First, they employed the rhetoric of biblical literalism in an age when the literal truth of the Bible was no longer the consensus.

86. Aviezer's first book has been translated into Hebrew and at least six European languages. Schroeder's *GBB* has been translated into at least five and *SoG* into at least four.

87. *American Jewish Identity Survey, 2001: An Exploration in the Demography and Outlook of a People* (New York: Graduate Center of the City University of New York, 2001), 20.

88. Ibid., 35–38.

89. *National Jewish Population Survey, 2000–2001* (New York: United Jewish Communities, 2001), 12.

90. Ibid., 10, 5; *Highlights of the CJS 1990 National Population Survey*, 32.

Second, they claimed to read Scripture literally, whereas the rabbinic tradition never privileged a literal reading, especially of the Creation narrative. Third, they focused on one aggadic passage, while ignoring incompatible aggadic accounts of Creation. In addition to these methods, their writings manifest the primary motives of fundamentalism: To defend God, the divinity of the Torah, and their own community's way of life.

Post-Script(ural) Anachronism

Rashi, the medieval biblical commentator par excellence, began his commentary on Genesis by asking, "What's this book of stories doing in the Torah?" What kind of a question is that? Wasn't Rashi just as interested in questions of origins as we are? Whence the world? Whence the different kinds of plants, animals, and peoples? And whence this particular people who is expected to do what is just and right, and about whom the nations of the world will say, "Surely, that great nation is a wise and discerning people" (Genesis 18:19, Deuteronomy 4:6)?

After the Second Temple was destroyed in 70 CE, the rabbis found refuge in halakhah. The House of God, the Temple in Jerusalem, had been razed by the Romans; but for rabbinic Judaism, God was now to be found in the indestructible Torah and the laws for living based on Torah study.[91] In the name of Ula, Rabbi Hiya son of Ami said: "Since the day the Temple was destroyed, the Holy One has in this world nothing other than the four cubits of halakhah."[92] To be sure, not all agreed with Ula. But in Rashi's opening comment on Genesis, we hear Ula's echo: "If God is found in the halakhah, why do we need the stories of Genesis?"

Our Modern Orthodox physicists might answer:

> Ula was right. To know God's will, study halakhah. But today, people (especially Jews!) don't believe in God at all. By demonstrating the concordance between the description of Creation in Genesis and contemporary science, we will draw people back to the belief in God and in the divinity of the Torah. Then let them study the halakhah. Rashi, that's why Genesis was included in the Torah—as a tool for outreach!

Our Modern Orthodox fundamentalists read God back into Creation—Creation as described in the first chapter of Genesis—in order to reenchant the world and to reverse the disenchantment of modern Jews from the beliefs and practices of traditional Judaism.

91. Avot de-Rabbi Natan, ch. 4.

92. Berakhot 8a.

8

Torah and Madda? Evolution in the Jewish Educational Context

Rena Selya

Education has long been a flashpoint in the ongoing public debate over Darwinian evolution in the United States. American high school biology classrooms and textbooks have been the focus of community concern throughout the twentieth century. From the Scopes trial in 1925 to contemporary attempts to legislate equal time for theories of Intelligent Design in public school curricula, the teaching of evolution offers a window onto any community's attitudes towards science, the Bible, and man's place in the world.[1] To some with religious objections to evolution, teenagers seem particularly vulnerable to corruption by this branch of science.

Jewish teenagers may be an exception to this rule. Despite the perception among many Orthodox Jews that evolution is fundamentally incompatible with a Torah perspective, contemporary Jewish thinkers have found many ways to reconcile the two.[2] In Modern Orthodox yeshiva high schools across the United States and Canada, Jewish teenagers study evolution, often with, but sometimes without, discussion of the religious implications and possible conflicts. It is the content of those religious discussions (or their absence) that is indicative of contemporary Jewish intellectual attitudes towards science, education, and biblical interpretation.

I am grateful to Yisroel Brumer, Geoffrey Cantor, Carl Feit, Joel Gereboff, Yirmiyahu Luchins, Tamar Kaplan, Ronald Numbers, and Marc Swetlitz for their helpful comments and suggestions, and to Barbara Selya for her skillful editing.

1. On the Scopes trial, see Edward J. Larson, *Summer for the Gods: The Scopes Trial and America's Continuing Debate over Science and Religion* (Cambridge: Harvard University Press, 1997). See Ronald L. Numbers, *The Creationists* (New York: Knopf, 1992), for a historical introduction to creationism. On more recent controversies in education see Larry A. Witham, *Where Darwin Meets the Bible: Creationists and Evolutionists in America* (Oxford: Oxford University Press, 2002).
2. For a discussion of these attempts, see Cherry (chapter 7) in this volume.

Within the umbrella of Orthodox Judaism, there is a spectrum of religious attitudes and practices. Sociologists Samuel Heilman and Steven Cohen have described the differences between "right wing," "traditionalist," or ultra-Orthodox Jews and their more "left wing," "centrist," or Modern Orthodox contemporaries as offering different responses to acculturation in American society. While the insular ultra-Orthodox tend to reject secular culture and adhere to strict observance of Jewish law, Modern Orthodox Jews have a more open attitude towards American culture and are relatively more lax in their religious practices.[3] Because ultra-Orthodox schools often omit secular subjects from their curricula, this chapter will focus on how Modern Orthodox high schools, also known as yeshiva high schools, approach the potentially controversial topic of evolution.

One of the defining characteristics of Modern Orthodoxy is its attempt to participate fully in the secular world, while maintaining a strong commitment to living according to the Torah and mitzvot (commandments). Although a person's Jewish identity and practice should be primary, Modern Orthodoxy accepts, even encourages, the individual to engage with general society in meaningful ways.[4] In Modern Orthodox circles, secular education, including science, is highly valued both as a vehicle for worldly success and as a way to enhance Judaism.[5] As evidence of their academic rigor, yeshiva high schools regularly publicize the number of graduates who have attended such prestigious universities as Harvard, Columbia, Yale, and the University of Pennsylvania.[6] Science, whether for its own sake or as a stepping stone to a career in medicine, is a key area that Modern Orthodox educators emphasize.

This paper will outline the American and Jewish contexts of science education in contemporary Modern Orthodox yeshiva high schools, and examine the published resources available to students and educators before describing how the topic of evolution is treated in biology classrooms in North American Jewish schools. These preliminary findings are based on interviews with teachers and administrators, as well as an e-mail survey of yeshiva high school

3. Samuel C. Heilman and Steven M. Cohen, *Cosmopolitans and Parochials: Modern Orthodox Jews in America* (Chicago: University of Chicago Press, 1989), 11–23.

4. For a discussion of the path Modern Orthodoxy has taken in the United States, see "The Sea Change in American Orthodox Judaism: A Symposium," ed. Hillel Goldberg, *Tradition* 37, no. 4 (1998).

5. Responses from Walter S. Wurzburger and Michael Wyschogrod in Goldberg, "Sea Change," 136–42.

6. See Jeffrey S. Gurock, "The Ramaz Version of American Orthodoxy," in *Ramaz: School, Community, Scholarship, and Orthodoxy*, ed. Jeffrey S. Gurock (New York: Ramaz School, 1989), 44, 73; Heilman and Cohen, *Cosmopolitans and Parochials*, 138.

alumni. This analysis is not a comprehensive survey of all Jewish high schools, but focuses mainly on five Modern Orthodox schools on the East Coast of the United States, with data from seven additional institutions. As a product of the yeshiva high school system and a member of a Modern Orthodox community, I relied on my personal and community networks to obtain this information.[7]

Although it focuses on a very specific issue, this chapter raises larger questions about the quality of Jewish education in contemporary American communities. There is a serious shortage of qualified science teachers in the United States, and Jewish educators face an ever bigger challenge in finding science teachers who are sufficiently well-educated in both Jewish thought and science to engage the challenges posed by evolution.[8] Two of my informants, Carl Feit and Jeremy Luchins, have Ph.D.s in scientific subjects as well as rabbinic ordination, and they teach both Jewish studies and biology to high school students in New Jersey. They both insist that a reconciliation between these topics requires a sophisticated understanding of both Torah *and* science, but they also recognize that they are atypical among science teachers in yeshiva high schools. Nevertheless, they both feel strongly that by the time their students graduate, they should be capable of studying and understanding the account of Creation in the Bible using both traditional and contemporary approaches that do not rely on a literal interpretation. In addition, these graduates should be able to study modern biology at a high level. As representatives of the schools that make a conscious effort to address perceived conflict between evolution and Judaism, they expect their students to have the skills to confront any intellectual or theological problems that arise.

This view is not universal, however. A recent pamphlet warning parents about the dangers of secular university life facing Modern Orthodox youth specifically identified evolution and biblical interpretation as potentially corrupting. In the spring of 2003, Gil Perl and Yaacov Weinstein, graduate students at Harvard and MIT respectively, wrote and distributed a pamphlet entitled *A Parent's Guide to Orthodox Assimilation on University Campuses.*[9] Based

7. I attended the Cincinnati Hebrew Day School for elementary school, and graduated from the Frisch School in 1991.

8. On the shortage of qualified science teachers in the United States, see Division of Research, Evaluation and Communication, Directorate for Education and Human Resources, *The Learning Curve: What We Are Discovering about U.S. Science and Mathematics Education*, ed. Larry E. Suter (Washington: National Science Foundation, 1996 (NSF 96–53)), and reports on science education from the National Research Council, available at http://www.nas.edu/nrc (accessed 10 January 2004).

9. Gil Perl and Yaacov Weinstein, *A Parent's Guide to Orthodox Assimilation on University Campuses*, available at http://www.rabbis.org/secular.htm (accessed 10 January 2004).

on their experiences as teaching fellows at these two universities, they argued that, in offering a range of ideas and ideologies, secular universities are dangerous places for Orthodox youth. Perl and Weinstein do not share the faith of some educators who accept that high school students possess the ability to grapple with difficult concepts. Instead, Perl and Weinstein do not see any religious value in teaching high school students "potentially troubling theories such as evolution," even with "Orthodox responses to the challenges they pose."[10]

The debate over evolution and Judaism in Orthodox schools thus reflects larger educational concerns in the Modern Orthodox community. American Jewish historian Jonathan Sarna points out that "schools serve as the primary setting . . . where American Jews confront the most fundamental question of American Jewish life: how to live in two worlds at once, how to be both American and Jewish."[11] For Modern Orthodox Jews in particular, this "central drama of American Jewish life is introduced and rehearsed" in high schools,[12] but the implications resonate far beyond the high school walls. The treatment of evolution, a topic that requires intellectual and religious sophistication, can be an indication of an Orthodox community's religious values and social priorities.

Evolution, Creation, and Education in the United States

American high school students encounter evolution by natural selection, species change, and the struggle for survival as part of the standard science curriculum. Biological ideas are, of course, both common and important in American culture, and schools are an essential means for imparting cultural knowledge. As Philip Pauly argues in *Biologists and the Promise of American Life,* in the early twentieth century, textbook authors and curriculum designers explicitly constructed biology courses "to produce Americans who would be modern, secular and humanistic."[13] An evolutionary perspective on disease and behavior was crucial to the progressive social agenda of the biology educators Pauly focuses on. Their textbooks, which contained relatively little about descent, but maintained an evolutionary worldview in natural history and geology, were standard fare from the 1920s on.[14] Education professor

10. Ibid., 6.
11. Jonathan D. Sarna, "American Jewish Education in Historical Perspective," *Journal of Jewish Education* 64, no. 1–2 (winter–spring 1998): 8–21, on 9.
12. Ibid., 10.
13. Philip J. Pauly, *Biologists and the Promise of American Life: From Meriwether Lewis to Alfred Kinsey* (Princeton: Princeton University Press, 2000), 172.
14. Ibid., 191–92.

Gerald Skoog has undertaken several statistical analyses of the topic of evolution in high school biology textbooks. He examined over ninety textbooks published between 1900 and 1977, documenting the number of lines and pages each devoted to a range of evolutionary topics. He found a steady increase in the emphasis on evolution until 1950, and then a sharp increase in the 1960s, with the publication of the textbooks in the Biological Sciences Curriculum Study (BSCS) series.[15] After a dip in the 1970s, biology texts in the 1990s became even more oriented towards evolution.[16] Especially after the BSCS textbooks made evolution a unifying theme in biology curricula, it was impossible to avoid at least some mention of Darwin in biology classes.

Evolution, then, is a commonly found concept in high school and college biology classrooms, and with good reason. As Theodosius Dobzhansky famously remarked, "Nothing in biology makes sense except in the light of evolution."[17] Since the evolutionary synthesis of the 1940s, it has been a cornerstone of experimental and theoretical work in the life sciences.[18] Students in college preparatory programs, whether in public or private schools, will encounter the ideas of natural selection, competition, and speciation.[19] One quarter of the Advanced Placement (AP) biology course is devoted to heredity and evolution, and evolution is also discussed in the parts of the course focusing on organisms and populations.[20]

Despite the importance of evolution in scientific practice and education, it remains a contentious topic in American culture. Creationists offer a range of critiques of evolution, from the argument that it is "just a theory," so it should not be taken seriously, to the claim that evolution has never been observed, so it cannot be true, to calculations of the probability that the mutation rate could cause species change. They also assert that the world

15. Gerald Skoog, "The Topic of Evolution in Secondary School Biology Textbooks: 1900–1977," *Science Education* 63 (1979): 621–40, reprinted in *Evolution versus Creationism: The Public Education Controversy,* ed. J. Peter Zetterberg (Phoenix: Oryx Press, 1983), 65–89. See also Witham, *Where Darwin Meets the Bible,* 151, and references to ch. 8.

16. Witham, *Where Darwin Meets the Bible,* 159.

17. Theodosius Dobzhansky, "Nothing in Biology Makes Sense Except in the Light of Evolution," *American Biology Teacher* 35 (1973), reprinted in Zetterberg, *Evolution versus Creationism,* 18–28.

18. Ernst Mayr and William Provine, eds., *The Evolutionary Synthesis: Perspectives on the Unification of Biology* (Cambridge: Harvard University Press, 1980); Vassiliki Betty Smocovitis, *Unifying Biology: The Evolutionary Synthesis and Evolutionary Biology* (Princeton: Princeton University Press, 1996).

19. See Pauly, *Biologists and the Promise of American Life,* ch. 7.

20. College Board, "Course Description: Biology," New York: College Entrance Examination Board, 2003, 5–14; downloaded from College Board Web site (http://apcentral.collegeboard.com/article/0,3045,151–165–0–2117,00.html), 6 January 2004.

itself is evidence of an intelligent creator and offer quasi-scientific evidence for divinely ordained Creation. The biblical account of Creation is not always cited as an alternative natural history, but some form of special creation of stable species by a divine being is always invoked. State legislatures and boards of education across the country regularly consider granting equal time to a creationist perspective, despite the legal proscription against the teaching of religious ideas in public schools.[21] In 2005, the issue again made headlines, with calls by President Bush and others to "teach the controversy" and to include Intelligent Design in public school science classrooms. In a high-profile court challenge brought by parents in Dover, Pennsylvania, the judge found that Intelligent Design is not science and "cannot uncouple itself from its creationist, and thus religious, antecedents."[22]

Modern Orthodox Jewish Education in America

Since World War II there has been a rapid increase in the number of Jewish elementary and high schools in the United States. Most of these schools are Orthodox, although other denominations are well represented. These day schools, as they are known, provide religious and secular education, as well as social and cultural opportunities for Jewish children. Many of these schools choose to join such organizations as Torah Umesorah, an umbrella organization that provides educational materials and sample Judaic curricula, but does not set guidelines for the content of secular studies.[23]

Day schools—especially Modern Orthodox ones—are expected to prepare their children for attending competitive colleges and universities, thus enabling them to participate extensively in the wider American society. At the same time, these schools are responsible for ensuring that their students will remain active, committed Jews when they proceed to secular universities.[24]

21. See selections in Zetterberg, *Evolution versus Creationism*, for a dated but comprehensive discussion of the legal and social issues in education; also Witham, *Where Darwin Meets the Bible*, ch. 8.

22. Laurie Goodstein, "Issuing Rebuke, Judge Rejects Teaching of Intelligent Design," *New York Times*, 21 December 2005, A1; "Excerpt from the Ruling on Intelligent Design," ibid., A34.

23. Doniel Zvi Kramer, *The Day Schools and Torah Umesorah: The Seeding of Traditional Judaism in America* (New York: Yeshiva University Press, 1984).

24. On the history and sociology of Jewish education, see Lloyd P. Gartner, "Jewish Education in the United States," in *The Jewish Community in America*, ed. Marshall Sklare (New York: Behrman House, 1974), 221–48; Harold S. Himmelfarb and Sergio DellaPergola, eds., *Jewish Education Worldwide: Cross-Cultural Perspectives* (New York: University Press of America, 1989); Samuel C. Heilman, *Portrait of American Jews: The Last Half of the 20th Century* (Seattle: University of Washington Press, 1995); Etan Diamond, *And I Will Dwell in Their Midst: Orthodox Jews in Suburbia* (Chapel Hill: University of North Carolina Press, 2000), ch. 4.

The dual curriculum is a reflection of the ideological position of the Modern Orthodox movement to participate in secular culture while maintaining a strong and visible Jewish identity. The world is seen through the lens of Torah Judaism, but the whole world is open for examination. There are many Modern Orthodox medical doctors and scientists who have trained at top universities.[25]

This ideology is also the source of some tension in the Modern Orthodox community. Secular culture—from philosophy and science to literature and movies—can be seen as directly opposed to Jewish values. Some have therefore felt that Jewish children should not be exposed to these areas too early, or possibly at all. During the development of the Torah Umesorah movement in the 1950s, there was interest in rewriting all of the general studies textbooks for Jewish elementary schools, including English readers and social studies and science books, because "they stated as facts many hypotheses that were antithetical to Jewish beliefs."[26] Torah Umesorah did not pursue the project, but nevertheless struggled to define guidelines for competitive general studies programs that would not compromise the Jewish faith of their students. In 1969 Rabbi Moshe Feinstein, an influential leader of American Orthodoxy, even advocated tearing out pages of science textbooks that contained references to evolution or other "matters of heresy."[27]

As we shall explore below, over the past fifteen to twenty years, there has been a push from moderate members of the Modern Orthodox community to articulate the ways in which science and Torah are compatible. For some educators, science is so readily integrated into the Jewish perspective that it is not even worth spending classroom time discussing possible conflicts between the two.[28] In ideal situations, teachers such as Carl Feit and Jeremy Luchins serve on both the Judaic studies and science teaching staff, and so are able to guide their students through the scientific and religious discussions of evolution. However, the difficulty in finding qualified science teachers, coupled with the compartmentalization of spiritual and religious education in Jewish schools, makes this kind of integration a rare occurrence.[29]

25. Heilman, *Portrait,* 38–41, 80.

26. Kramer, *Day Schools and Torah Umesorah,* 67–68.

27. Moshe Feinstein, *Igrot Moshe, Yoreh De'ah,* vol. 3, responsum 73 (New York: Noble Press, 5742 [1982]), 323.

28. Conversations with Rabbi Menachem Meier, 1 November 2003, and Kalman Stein, 13 January 2004.

29. On the compartmentalization, see Katherine G. Simon, "Bring It Up with the Rabbi: The Specialization of Moral and Spiritual Education in a Jewish High School," *Journal of Jewish Education* 64, no. 1–2 (winter–spring 1998): 33–43.

There is some evidence that, in the educational context, science has become less of a threat, while literature and music have become more problematic. In a defense of teaching secular studies, Rabbi Mayer Schiller refers to his experience teaching in a Modern Orthodox high school, and argues that "the great crisis which confronts Modern Orthodoxy [is] hedonism, not ideology. . . . No, it is not necessary to throw out our volumes of classical poetry or great music; it is merely necessary to smash the television and shatter junior's CD collection."[30] Similarly, Rabbi Mark Gottlieb, principal of the Maimonides School in Brookline, claims that parents never object to teaching evolution, but they often have problems with the choice of literature that the students read in English class. *The Catcher in the Rye* and *A Thousand Acres* are much more controversial than science.[31]

College Models: Torah u-Madda at Yeshiva University, Creationism at Touro

In the late twentieth century, the engagement with Jewish and secular subjects has fallen under the rubric of "Torah u-madda" (Torah and worldly wisdom) as articulated by Norman Lamm, former president of Yeshiva University (YU).[32] In Lamm's view, "the synergistic interrelation of religious study and secular or profane knowledge . . . yields constructive results" for both Jewish and secular knowledge.[33] Science can have great moral authority, and provide practitioners with the power to improve the world. If Torah and secular knowledge somehow conflict, the Torah perspective prevails. In this model, there are no religious issues with studying Torah and the natural sciences in general, and evolution in particular, as long as one has the proper understanding of Torah. "The various theories of evolution and geology pose serious problems for those who have not yet resolved for themselves a proper reading of Genesis, as proposed by sophisticated Torah sages of recent generations who met the problem head-on without resorting to simple denial."[34]

30. Mayer Schiller, "*Torah Umadda* and *The Jewish Observer* Critique: Towards a Clarification of the Issues," *Torah u-Madda Journal* 6 (1995–96): 58–90, on 76–77.
31. Interview with Rabbi Mark Gottleib, 8 January 2004.
32. Since the establishment of Yeshiva College in 1928, YU has offered a dual curriculum in Torah and secular studies, but in the 1970s, there was a need to formally define what was meant by the Torah u-madda model. Norman Lamm, *Torah Umadda: The Encounter of Religious Learning and Worldly Knowledge in the Jewish Tradition* (Northvale, NJ: Aronson, 1990), pp. xii, 10–12.
33. Ibid., 6.
34. Ibid., 206.

The "proper reading of Genesis" seems to be more of a contentious issue for many Jewish thinkers than the problematic philosophical implications of evolution. The religious philosophy behind Torah u-madda is not merely about the value of non-Jewish knowledge. The movement reflects a nonliteral approach to biblical interpretation as well as a commitment to participation in modern society and culture. This perspective, although widespread in the Modern Orthodox community, caused some controversy when Lamm's book first appeared in 1990.[35]

Scientific research and teaching are visible components of the intellectual life at YU, where we witness the most explicit application of the Torah u-madda perspective. Recently, the *YU Review,* the magazine for YU alumni, devoted an entire issue to "Science and the Global Community."[36] In an introductory essay, university president Richard Joel commented that "while our emphasis on scientific successes crosses all academic levels, the research we support has a moral underpinning." He identified that moral charge with "our *raison d'etre*—to marry the wisdom of faith with the need to explore our universe's mysteries."[37] An article in the same issue of the *YU Review* provocatively titled "Torah and Big Bang" described biology professor Carl Feit's ongoing intellectual project to unite scientific and religious considerations of the universe and its origins. He has also stated that no contradiction exists between the Jewish tradition and biology and argued that the evolutionary ideas that some find threatening can actually serve to strengthen one's faith.[38]

Feit reports that he devotes some time in his introductory biology class to explaining to his students why the study and acceptance of evolution by natural selection are unproblematic. In order to make an informed decision about the issues his students read sources on the philosophy of science, selections from Darwin and modern evolutionists, Jewish biblical commentaries, and works by philosophers.[39] Although he does not alter the content of his biology syllabus in any way, the inclusion of Jewish sources in this college biology course suggests that some students might otherwise be disturbed by the subject.

35. Schiller, "*Torah Umadda,*" 58–90.

36. Special issue devoted to "Science and the Global Community," *YU Review* (summer 2003).

37. Richard Joel, "A Journey Worth Taking," ibid., 3.

38. Norman Eisenberg, "Torah and Big Bang," ibid., 13–15. See also Robinson (chapter 3) and Feit (chapter 9) in this volume and Carl Feit, "Darwin and Drash: The Interplay of Torah and Biology," *Torah u-Madda Journal* 2 (1990): 25–36.

39. Conversation with Carl Feit, 9 March 2003.

Yeshiva University's explicit commitment to science sets it apart from other Orthodox Jewish institutions of higher learning in America. In a recent article, Alexander Nussbaum described his experiences teaching experimental psychology and statistics at Touro College, which was founded in 1971 in order "to enrich the Jewish heritage."[40] Touro tends to attract ultra-Orthodox students, in contrast to YU's more Modern Orthodox population. An offhand remark of Nussbaum's in support of teaching evolution elicited exclamations of "horror, curiosity and surprise" from his students.[41] Upon further discussion, he found that other science professors at Touro routinely taught creationism and criticized evolution. He argues that while ultra-Orthodox institutions like Touro "crave, claim and take pride in the prestige of scientific academia" and go to great lengths to achieve it, they nevertheless mock and dismiss this crucial aspect of modern scientific thought.[42] Evolution is beyond the bounds of "safe" scientific ideas for this ultra-Orthodox institution. This episode illustrates some of the tensions inherent in the Orthodox community's attempts to embrace and contain scientific ideas simultaneously.

Literature on Judaism and Evolution

Contributions to this volume by Robinson (chapter 3) and Cherry (chapter 7) provide detailed analyses of the literature on Judaism and evolution from an Orthodox perspective.[43] In the early 1970s and '80s, the Association of Orthodox Jewish Scientists and a group called B'or Ha'Torah (By the Light of the Torah) offered commentary on science and Judaism for scientists, students, and laymen. These works by and large repeat the creationist critiques offered by Rabbi Menachem Mendel Schneersohn, the Lubavitcher rebbe, that ridicule evolutionary theory and justify a literal interpretation of Genesis.[44] Interestingly, these attempts to explain away evolution were written at a time when Orthodox Jews seemed to have little quarrel with their

40. Alexander Nussbaum, "Creationism and Geocentrism among Orthodox Jewish Scientists," *National Center for Science Education Reports* 22, no. 1–2 (Jan.–Apr. 2002): 38–43. I am grateful to Ronald Numbers for pointing me to this source. Information about Touro's mission from its Web site, www.touro.edu (accessed 5 January 2004).

41. Nussbaum, "Creationism," 38.

42. Ibid., 42.

43. See contributions to this volume by Robinson and Cherry, as well as Cherry, "Three Twentieth-Century Jewish Responses to Evolutionary Theory," *Aleph: Historical Studies in Science and Judaism* 3 (2003): 247–90.

44. Aryeh Carmell and Cyril Domb, eds., *Challenge: Torah Views on Science and Its Problems* (London: Association of Orthodox Jewish Scientists; Jerusalem: Feldheim, 1976); H. Chaim

children studying evolution. In a survey of Orthodox Jews in the New York metropolitan area conducted in 1979 and 1980, sociologists Samuel Heilman and Steven Cohen found that a "vast majority" agreed with the statement "My children should learn about the theory of evolution in school."[45]

In contrast to these earlier sources, three books and articles written in the 1990s by Carl Feit, Judah Landa, and Baruch Sterman, all Orthodox Jews and practicing scientists, offer Jewish perspectives that allow for, or even insist on, an evolutionary worldview. Rather than seeing science as a threat to Jewish beliefs, they argue that it can and even should be a part of an Orthodox belief system.[46]

All three authors are at the more liberal end of the spectrum of Modern Orthodoxy. It is interesting to note that these articles advocating a more open stance towards evolution and a move away from biblical literalism appeared at a time when the Modern Orthodox community was becoming stricter in religious practice. Despite this change in observance, these arguments in favor of reconciling Torah and evolution evoked little or no controversy.[47] This trend contrasts sharply with the situation in the 1970s and '80s, when Modern Orthodox sources generally portrayed science and religion as fundamentally incompatible.

Literature on Teaching Evolution in Jewish Contexts

In addition to these general reflections on Darwin and Torah, there are also a handful of sources explicitly addressing the question of how to teach evolution in Jewish schools. These works reflect the historical shift in the general literature on Judaism and evolution. Earlier works tended to explain evolution away, while later perspectives urged a balanced view of the separate

Schimmel and Aryeh Carmell, eds., *Encounter: Essays on Torah and Modern Life* (Jerusalem: Association of Orthodox Jewish Scientists/Feldheim, 1989), esp. contributions by Alvin Radkowsky, "Miracles," 42–74, and Herman Branover, "Basic Principles in the Discussion of Torah and Science," 232–42; and Herman Branover and Ilana Coven Attia, eds., *Science in the Light of Torah: A B'or Ha'Torah Reader* (Northvale, NJ: Aronson, 1994).

45. Heilman and Cohen, *Cosmopolitans and Parochials,* 158–60.

46. Feit, "Darwin and Drash"; Judah Landa, *Torah and Science* (Hoboken, NJ: Ktav, 1991); Baruch Sterman, "Judaism and Darwinian Evolution," *Tradition* 29 (1994): 48–75.

47. The two letters (from Gil Melmed, a University of Pennsylvania undergraduate, and Rabbi Zvi Grumet, a yeshiva high school teacher, *Tradition* 29 (1995): 86–88) to the editor of *Tradition* in response to Sterman's article were both positive, thanking the editors for publishing such a thoughtful and helpful article. For a brief analysis of Modern Orthodoxy's turn to the right, see Heilman, *Portrait,* 144–52, as well as his contribution to Goldberg, "Sea Change," 77–80.

spheres of influence of science and of Torah. They cautioned students not to look to science as a religious source, or at the Torah as a scientific text, but rather to see each as a legitimate authority on different aspects of modern life.

In 1978, the Bureau of Jewish Education (BJE) in New York published *Viewpoints on Science and Judaism,* a collection of twenty-six essays edited by Tina Levitan. Alvin Schiff, the executive vice president of the BJE, explained that "teachers, students and parents are frequently beset by questions of ethics, belief and behavior that result from the differing moods and contrasting operational premises of science and Judaism."[48] The BJE therefore produced a resource that brought together the perspectives of active scientists from various sections of the Jewish community as an educational tool for promoting discussion. According to Levitan, "That science and religion do not clash with one another is clear to those who engage in careful and diligent study of the Torah."[49] She surveyed American Jewish scientists from a range of disciplines to see whether and how participation in science affected their religious beliefs. Not all of the scientists included in this volume were religious, and she also included selections from five philosophers of science and religion.

The general thrust of these essays and interviews is that there is no conflict between science and Judaism. Most of the scientists cited claim that scientific knowledge strengthens their religious faith, since physics, astronomy, biochemistry, and medicine all reveal the glory of God through the lawlike behavior of his creation. However, the essay on evolution by natural selection by Morris Goldman repeated the creationist critique he contributed to the volume published by the Association of Orthodox Jewish Scientists. He argued that the Torah and science give two very different accounts of Creation, and since we cannot scientifically account for any divine origins of the world, "we must, therefore, resort to the indirect approach of seeing how well or how poorly the Darwinian doctrine conforms to the requirements of a scientific theory."[50] He then presented the familiar "scientific" arguments against Darwinian evolution: the difficulty of defining fitness, lack of reproducibility of evolutionary change in a laboratory, the fact that no speciation has been observed either in the laboratory or in the wild, and the inadequacy of the fossil record. He concluded by

48. Alvin I. Schiff, foreword to *Viewpoints on Science and Judaism*, ed. Tina Levitan (New York: Bureau of Jewish Education, 1978), p. vii.

49. Tina Levitan, introduction, ibid., p. xi.

50. Morris Goldman, "Evolution by Natural Selection," ibid., 48–54, 50; Morris Goldman, "A Critical View of Evolution," in Carmell and Domb, *Challenge*, 216–34. A Hebrew version appeared in Yaacov Kornreich, ed., *NCSY Torah and Science Reader* (New York: Orthodox Union, 1971).

reassuring the reader that "[t]he true story of life in scientific terms is yet to be told. If 4,000 years of Jewish history is any guide, no believing Jew need feel any anxiety about how his faith will fare when the full story is known."[51]

Because youth groups are an important locus for formal and informal Jewish education, I contacted the largest Orthodox youth group, National Conference of Synagogue Youth (NCSY), to see if it had any educational materials on evolution.[52] In 1971, NCSY published a Hebrew reader with sources on science and Torah that presented the creationist arguments offered by Goldman and others,[53] but that text does not seem to be widely used. Rabbi Jack Abramowitz, the director of national programs at NCSY, assured me that there is no conflict between Judaism and evolution, and so it is not a "hot button" issue for participants. NCSY had not developed any materials on evolution, but he referred me to Josh Greenberger's *Human Intelligence Gone Ape*, published by NCSY and its parent organization, the Orthodox Union (OU), which they use as a resource.[54] Surprisingly, this book is an extreme example of creationist rhetoric with very little Jewish content. Greenberger, a self-educated computer consultant and "popular columnist," sarcastically offered the usual pseudoscientific arguments against evolution, while also suggesting that the emergence of AIDS in the 1980s provided evidence of divine creation and continued intervention in the world.[55] No one currently at the OU was able to explain how or why they had published this text.

In a short 1997 article in *Ten Da'at: A Journal of Jewish Education,* Joel Wolowelsky argued that evolution had to be a key component in any successful yeshiva high school curriculum. Drawing on his experiences as chairman of advanced placement studies at Yeshiva of Flatbush, a coeducational yeshiva high school, Wolowelsky observes that some Jewish high schools either skip evolution altogether or else degrade evolution by reminding students that it is "just a theory." He claims that "these approaches are bad *hashkafah* [Jewish outlook] and worse science." Wolowelsky accepts that there are parts of Genesis that may give an alternate narrative of the history of the world, but argues that this narrative has no place in the science classroom, since "the Torah is an eternal book of truth; as such it cannot be a sci-

51. Goldman, "Evolution by Natural Selection," 54.

52. For an analysis of the role of NCSY, see Nathalie Friedman, *Faithful Youth: A Study of the National Conference of Synagogue Youth* (New York: OU/NCSY, 1998).

53. Kornreich, *NCSY Torah.*

54. Conversation with Jack Abramowitz, 5 January 2004; Josh Greenberger, *Human Intelligence Gone Ape* (New York: OU/NCSY, 1990). This is basically the same text as his *Theories and Fantasies: Evolution: Force or Farce?* (New York: Systematic Science, 1986).

55. Greenberger, *Human Intelligence,* 75–80.

ence textbook." He is convinced that with a proper introduction, even a topic as controversial as evolution will ultimately lead students to bless God for his "hidden guiding hand." He concludes that "there is no contradiction between good science and Torah *hashkafah*." Wolowelsky feels that any apparent contradiction between science and Torah is based on "a shallow understanding of Torah or an unsophisticated appreciation of science—or both." As an educator, he is convinced that if students have the tools to appreciate both science and Torah in depth, they will conclude that there is no problem with evolution.[56]

Although they were intended for Jewish educators and teenagers, these materials do not seem to have been used in the science classrooms of yeshiva high schools. Most of the rabbis who visit the science classes in order to show the pupils how to reconcile science and Judaism use many of the biblical and talmudic sources that Feit and Sterman cite. The published literature seems to be used mainly by other academics or practitioners of science, as opposed to educators teaching in Jewish high schools.[57]

How Jewish Schools Teach Evolution

In order to get a sense of how Darwinian evolution is treated, I contacted teachers and administrators at five large, well-established yeshiva high schools on the East Coast of North America. I received information and materials from Rabbi Menachem Meier, former principal of the Frisch School; Dr. Kalman Stein, the current principal of Frisch; Rabbi Jay Goldmintz, headmaster of the Ramaz Upper School; Rabbi Mark Gottleib, the principal of the Maimonides School and a former teacher at the Ida Crown Jewish Academy; Rabbi Jack Bieler, formerly at the Melvin J. Berman Hebrew Academy; and Rabbi Dr. Jeremy Luchins, chair of the science department at the Joseph Kushner Academy.[58] I also sent an e-mail query to members of the Orthodox Minyan in Cambridge, Massachusetts, a community made up of Harvard undergraduates, graduate students at several Boston-area universities, and residents of the

56. Joel Wolowelsky, "Teaching Evolution in Yeshiva High School," *Ten Da'at: A Journal of Jewish Education* 10, no. 1 (spring 1997): 33–39, see http://www.daat.ac.il/daat/english/education/evolution-1.htm (accessed 5 January 2004).

57. See, for example, the two letters cited in n. 47.

58. Joel Wolowelsky of Flatbush and Rabbi William Altschul from the Hebrew Academy in Washington did not respond to my queries, but see the discussion below of the program that Rabbi Jack Bieler ran at the Hebrew Academy.

YESHIVA HIGH SCHOOL PERSPECTIVES ON EVOLUTION

Name	*Location*	*Coed?*	*Evolution in the classroom?*	*Religious discussion?*	*Religiously compatible?*
Ramaz	New York	yes	yes	yes	yes
Frisch School	Paramus, NJ	yes	yes	sometimes	yes
Joseph Kushner Hebrew Academy	Livingston, NJ	yes	yes	yes	yes
Maimonides School	Brookline, MA	yes	yes	sometimes	yes
Melvin J. Berman Hebrew Academy	Rockville, MD	yes	yes	yes	yes
Yeshiva of Flatbush	Brooklyn	yes	yes	sometimes	yes
Ida Crown Jewish Academy	Chicago	yes	yes	sometimes	yes
Stella K. Abraham High School	Hewlett Bay Park, NY	girls	yes	no	no
Shulamith School	Brooklyn	girls	no	no	no
New England Academy of Torah	Providence, RI	girls*	yes	yes	yes
Yeshiva B'nei Akiva Or Chaim	Toronto	boys	no	yes	no
Yeshiva University Los Angeles (YULA)	Los Angeles	boys*	yes	yes	no

* These schools have divisions for both boys and girls, but I only heard from alumni of the indicated sex.

town who participate in Orthodox services at the Harvard Hillel.[59] I received eighteen e-mail responses, as well as several answers in personal conversations with alumni of twelve Orthodox high schools. The results are summarized in the table. The comments from alumni will be incorporated into the descriptions of how evolution is taught at each of the five schools profiled.

59. The text of the e-mail sent to the or-minyan-misc list was: "I am doing research for a paper on how Jewish schools teach evolution, and was hoping to get some information from some graduates of those schools. If you went to a Jewish high school and remember your biology class, I would really appreciate hearing about your experiences (even if you went to high school a while ago). I am writing a paper on this topic for a conference on 'The Jewish Tradition and the Challenge of Evolution.'

Jewish schools adopt four approaches to teaching evolution. In some schools, biology teachers teach evolution without comment or question. In others, biology teachers teach evolution, and then a rabbi or other religious studies teacher comes into the science classroom and presents a Jewish perspective on the topic. That Jewish perspective could either be a way to read the Creation narrative in a nonliteral way, or to present an argument for intelligent design. The third option is to assign chapters from the book for students to read at home, without discussing these readings in the classroom. The fourth option is to not teach evolution at all. There is anecdotal evidence that some administrators rip the chapter out of science textbooks, but I did not encounter any such cases in my survey, nor did I find any instances of a creationist science curriculum replacing the standard biology texts.[60]

As the table indicates, most of these schools teach evolution with some explanation about how it is compatible with Judaism. Schools that separate boys and girls tend to be more cautious about teaching evolution than the coeducational schools profiled. In general, single-sex schools tend to be more religiously conservative than coeducational schools, which may explain this trend. At Shulamith, one alumna recalls that evolution was not taught at all; students were simply told to read it on their own, since it appeared on the New York State Regents Examination.[61] Similarly, at the Stella K. Abraham High School for Girls, the students were told to learn evolution for the Regents Examination and then to forget it.[62] At Yeshiva University, Los Angeles, the topic was addressed at a general school meeting for all of the boys, when a rabbi attempted to disprove evolution scientifically by arguing for the statistical improbability of evolution and by claiming that fossils were placed on earth by God as a test of faith.[63] An alumnus of Yeshiva B'nei Akiva Or Chaim in Toronto recalls several occasions when rabbis made arguments from design in order to disprove evolution to students in grades seven and eight, while his high school biology teacher made it clear to the students that they were skipping an important chapter in the textbook.[64]

I am interested in if there was ever any discussion of a perceived conflict between Judaism and evolution and who led that discussion. Were Rabbis or other non-scientists involved in the discussion? Was the biology curriculum modified in any way? If yes, how? Was it ever an issue among the students? Discussed in the school paper? Were parents ever involved? Any information, recollections or references would be great."

60. Wolowelsky, "Teaching Evolution"; Landa, *Torah and Science*, 290.

61. E-mail from Tzipora Russ, 16 December 2003.

62. E-mail from Rachel Kohl, 15 December 2003.

63. E-mail from Mayer Bick, 16 December 2003.

64. E-mail from Yisroel Brumer, 16 December 2003.

Although there is variation in how evolution is treated in their biology classrooms, the five schools profiled all fit the same model of Modern Orthodox high schools. These schools are all coeducational, with strong college preparatory departments. Ramaz and Maimonides were established by prominent figures in the Jewish community and have a longstanding commitment to excellence in both Jewish and secular education. Both schools opened in 1937. Ramaz was founded by Rabbi Joseph Lookstein on the Upper East Side of Manhattan, while Maimonides was founded in Brookline, Massachusetts, by Rabbi Joseph B. Soloveitchik. Soloveitchik also played a role in the establishment of the Frisch School in 1972. The Melvin J. Berman Hebrew Academy (formerly the Hebrew Academy of Greater Washington), part of the Torah Umesorah movement, was founded in 1944. The Joseph Kushner Hebrew Academy (formerly the Hebrew Youth Academy) is over fifty years old and explicitly identifies with the Torah u-madda philosophy.[65] In these schools evolution formed a part of the biology curriculum for both ninth- and twelfth-grade students. At times rabbis would offer explanations of how evolution and Judaism were compatible, with relevant Jewish sources and close readings of Genesis. Indeed, the issue was just as likely to come up in a Bible or Jewish philosophy class as in biology.[66] The teachers seem to be the main impetus for any kind of discussion in the science classroom.

Although Meier claimed that it was a nonissue—"the conflict between science and religion is with the church, not the synagogue"—he recalled times during his career at Frisch when he explained to the freshmen that "it is possible to be Orthodox in deed and creed and believe in evolution," but with some modifications to account for God's role in designing the universe.[67] Carl Feit is currently invited to talk to the ninth-grade and AP biology classes at Frisch, but Stein reiterated that evolution is "not a hot-button issue" because "we are not fundamentalists."[68] Frisch alumni recalled that on several occasions, one of the rabbis would visit the nonhonors classes and present ways that evolution could be reconciled with Torah.[69] One graduate from the class of 1989 remembered her biology teacher presenting evolution

65. See essays in Gurock, *Ramaz*. For the history of the schools, see the Web sites: www.maimonides.org, www.jkha.org, www.mjbha.org (accessed December 2003 and January 2004).

66. Flatbush information from e-mail from Michael Kress, 16 December 2003, and conversation with Kim Sparer, 10 January 2004.

67. Conversation with Rabbi Menachem Meier, 1 November 2003. Meier was at Frisch from its founding in 1972 to 1997.

68. Conversation with Carl Feit, 9 March 2003, and Kalman Stein, 13 January 2004.

69. This is my recollection from my ninth-grade class, and this was confirmed in conversation with Tamar Kaplan.

as scientific fact, and then challenging them to reconcile it with Judaism. The students dutifully asked their Talmud teacher, who explained that since the sun and moon weren't created until the fourth day in the biblical account, one should not interpret the word "days" literally.[70]

At the Maimonides School and at the Ida Crown Jewish Academy, the issue is addressed in a similarly "unsystematic" way. When Gottleib taught at the Ida Crown Jewish Academy in Chicago, the biology teacher, who was also an Orthodox Jew, asked him to discuss the topic with the AP students. He wanted to "make the case that the two are not irreconcilable." Having read the essays on species history in *The Panda's Thumb* by Stephen Jay Gould, the students discussed the idea of perfectibility and design from a Jewish perspective with Gottleib.[71] At Maimonides, one of the rabbis gave a short presentation to the AP students if the science teacher requested it, but it was not a formal arrangement.[72] Maimonides alumni recalled that this was a general discussion about science and religion in which both the time issue and evolution were emphasized.[73]

In contrast to the other administrators, Goldmintz found that evolution has "always [been] an issue for the kids" throughout his twenty-two years at Ramaz. Although it was ignored for a while, in recent years he has visited both the ninth- and twelfth-grade biology classes, presenting them with different models of how to think about evolution from a Jewish perspective. He presents the whole spectrum of responses, from the Lubavitcher rebbe's rejection of evolution to ways of reading Genesis as an allegory, as well as the dangers inherent in each of these models. He also reminds the students that the Torah is not a history book. He goes into more detail with the seniors than with the freshmen, since he has found that ninth-graders are often too eager for an explicit answer, whereas the older students have more skills to grapple with the issues. Goldmintz teaches a class on Genesis to seniors and discusses with them the challenges of evolution.[74] However, some Ramaz alumni recalled no such discussions, either in science or in Judaic studies classes. They simply learned evolution, and if they had religious questions, they were not addressed at school.[75]

70. E-mail from Shari Lowin, 16 December 2003.

71. Stephen Jay Gould, *The Panda's Thumb: More Reflections in Natural History* (New York: Norton, 1980). This teacher puts particular emphasis on evolution, assigning Gould's books over the summer before the course begins. Conversation with Raichel Cohen, 6 December 2003.

72. Conversation with Rabbi Mark Gottleib, 8 January 2004. Rabbi Dovid Shapiro usually gives the presentation.

73. E-mails from Gregory Bunimovich and Adina Gerver, 16 and 17 December 2003.

74. Conversation with Rabbi Jay Goldmintz, 19 November 2003.

75. E-mail from Adam Sandler, 16 December 2003, and conversations with Brian Zuckerman and Danny Sadinoff, December 2003.

Ramaz, Frisch, and Maimonides are probably typical of how Modern Orthodox schools treat evolution in a Jewish context. Evolution is included in biology curricula, and any religious questions are dealt with by the rabbis, either on an ad hoc basis or else in short formal presentations. However, I encountered two instances of schools taking a more in-depth or interdisciplinary approach to the issues.

In 2004 the Melvin J. Berman Hebrew Academy in Rockville, Maryland, held an interdisciplinary "Creation Day." Rabbi Jack Bieler, a teacher with a longstanding interest in "integrating Judaic and general studies," organized a day exploring different aspects of the topic of Creation, as part of a year-long study of Genesis by the whole school. The theme of creativity thereby united all of the activities that included discussions, text study, music and drama performances, and art projects. The session on evolution began with students acting out a scene from *Inherit the Wind*, and then the whole high school split into groups with different faculty members and parents from the community to discuss ways of reconciling Judaism and evolution.[76]

At the Kushner School, Jeremy Luchins is a senior rebbe and chair of the science department. He regularly discusses evolution from a Jewish perspective with his science students.[77] In his course description he identifies evolution as one of the "topics unique to a yeshiva setting [that is] presented both to heighten awareness of the issues and to deepen appreciation of Torah and Halachah."[78] One recent graduate claimed that he "found little or no difference between what he [Luchins] taught me and what I learned in Science B-29 [introductory biology] here at Harvard." Luchins gives a detailed discussion of the textual issues in Genesis, pointing out that the Torah uses the word *bara* (to create) only three times in the narrative: in the very beginning, when animals are created, and then when man is created. The student recalled how Luchins taught

> that evolutionary biology can explain everything except for the original creation, the creation of life, and the creation of consciousness. And that the Torah is consistent with this, as Rashi says the story of creation is not chronological and others say it need not be taken literally. And this year in Science B-29 "The Evolution of Human Nature" I was taught nothing that contradicted this view.[79]

Classroom time is also devoted to examining both religious and scientific critiques of natural selection.

76. E-mails from Rabbi Jack Bieler, 8 and 11 January 2004.

77. E-mails from Ari Moskowitz, 15 December 2003, and Ellie Kaplow, 19 December 2003.

78. J. Luchins, "Honors Biology: Scope, Goals and Philosophy" (course description), fall 2003.

79. E-mail from Ari Moskowitz, 15 December 2003.

Luchins also teaches a special after-school class on science and Judaism in which he focuses on the implications of biology for Jewish thought, using Jewish and scientific sources to give the students an in-depth analysis of the issues. He tries to show how traditional biblical commentators and contemporary scientific thinkers offer startlingly similar versions of the creation of the world and man's place in it. Although it is not a requirement, all of the students in the high school are encouraged to attend the class in order to gain the skills to evaluate the relationship between Torah and science. Luchins also teaches similar classes to adults in synagogue settings, and argues passionately that Jews must not shy away from sophisticated textual analysis of Genesis or accept superficial accounts of Darwinian evolution.[80]

Conclusion

From this survey, it seems that in the educational context, there is little conflict between contemporary Modern Orthodox Judaism and Darwinian evolution. In many Jewish high schools, teenagers study the ideas and evidence of evolution by natural selection and are exposed to sophisticated biblical commentaries on the natural world. Jewish scientists, scholars, and educators have worked long and hard to provide high school students with intellectually stimulating and religiously satisfying perspectives that do not compromise either Torah or science. Parents do not object to their children studying evolution, and students do not report having their faith shaken by exposure to evolution.

Paradoxically, the increased acceptability of teaching evolution to Jewish teenagers over the past twenty years coincides with an increasing political turn to the right among Modern Orthodox Jews. This change in religious practice is characterized by stricter adherence to Jewish law (halakhah), increased attention to spirituality as opposed to intellectualism, and a tendency to rely on more right-wing authorities in questions of theology.[81] It is unclear whether there will be accompanying social and cultural shifts in attitudes towards secular learning. Even with the current acceptance of Darwinian evolution by Modern Orthodox thinkers, the debate over education, Torah values, and evolution continues. Torah and madda remain in conversation.

80. Conversations with Rabbi Jeremy Luchins, February and March 2004.

81. Many of the participants in the "Sea Change" symposium refer to this turn to the right while maintaining that Modern Orthodoxy is still different from ultra-Orthodoxy in outlook and practice; see reference in footnote 4.

9

Modern Orthodoxy and Evolution: The Models of Rabbi J. B. Soloveitchik and Rabbi A. I. Kook

Carl Feit

> Evolutionary theory, which is now achieving such world-wide acclaim, coincides with the lofty doctrines of kabbalah more than any other philosophical doctrine.
>
> RABBI ABRAHAM ISAAC HA-KOHEN KOOK, *Orot ha-Kodesh,* 2:565

> I have never been seriously troubled by the problem of the biblical doctrine of creation vis-à-vis the scientific doctrine of creation at both the cosmic and organic levels.
>
> RABBI JOSEPH B. SOLOVEITCHIK, *The Lonely Man of Faith,* 7

Two thinkers of the last century undisputedly represent the best models of Modern Orthodoxy. Their teachings and writings created the very framework of twentieth-century Modern Orthodox thought. These outstanding philosophers were Rabbi Abraham Isaac ha-Kohen Kook (1865–1935), the first Ashkenazi chief rabbi of Palestine, and Rabbi Joseph B. Soloveitchik (1903–93), who served as the rosh yeshiva (rabbinic leader) of Yeshiva University's Rabbi Isaac Elchanan Theological Seminary and professor of philosophy at the university for almost fifty years. Both men were widely acknowledged, by all segments of the Jewish community, as outstanding traditional Torah scholars, and yet concern for the interface of tradition with modernity was central and pivotal to their thought.

Kook was born in 1865 and attended traditional Eastern European yeshivot, including the famed Volozhin yeshiva, in what is now Belarus, before immigrating to Israel in 1904, where he became the chief rabbi of Tel Aviv/Jaffa. In 1921 he was appointed the first Ashkenazi chief rabbi of Palestine, and in the same year founded a school now known as Mercaz HaRav. He was a child prodigy and soon established his reputation as a talmudic scholar and *posek* (judge). However, he was also deeply influenced by kabbalah, and his writings include poetic, mystical musings on a wide variety of topics, ranging from what would now be called religious Zionism, to the nature of just societies and just wars, to the relationship of Jews and Judaism to the rest of world culture, history, and society. His influence has been strongest on the Orthodox community in Israel.

Soloveitchik was the scion of one of the most illustrious families in Mitnagdic Judaism, which stresses the intellectual aspects of religious experience, in contrast to Hasidic Judaism, which stresses emotional experience. His grandfather, Ḥayyim Soloveitchik, is credited with revolutionizing talmudic study by introducing an incisive method of analysis. Soloveitchik was himself a child prodigy in a family of geniuses and, aside from a brief early period of tutelage under a local Hasidic teacher, he spent the rest of his Jewish education studying Talmud with his father, Moses Soloveitchik. Largely due to the influence of his mother, Soloveitchik left Warsaw, where he had achieved the equivalent of a high school education, in order to enroll in the University of Berlin in 1926, where he spent five years studying philosophy, mathematics, and science. He wrote his Ph.D. thesis on the epistemology of the neo-Kantian philosopher Hermann Cohen. Immigrating to the United States in 1932, he spent the next ten years as a rabbi in Boston, where he was founding dean of the Maimonides School. In 1942 he succeeded his father as rosh yeshiva at Yeshiva University and continued to give advanced classes in Talmud until his retirement. He also periodically taught Jewish philosophy in the university's Bernard Revel Graduate School of Jewish Studies. He is credited with having ordained more rabbis than anybody else. His influence on American Modern (or centrist) Orthodoxy is immeasurable.

Although both Soloveitchik and Kook were remarkably creative writers, and were productive in many ways, neither published a large body of literature in his lifetime, although Kook's voluminous writings are being published posthumously.[1] In their writings they frequently alluded favorably to the accomplishments of modern science, and neither expressed any substantive problems with contemporary evolutionary theory. In this essay, I will show that their responses to early twentieth-century science were based on very traditional and normative Jewish sources. I will also explore the importance of cosmological and evolutionary motifs in their thinking.

1. To date, Kook's students have published but a small portion of his manuscripts, poems, and diaries. The quality of the editing varies tremendously. His halakhic decisions have been published separately as *Mishpat ha-Kohen* (Jerusalem: Mossad ha-Rav Kook, 1937).

Soloveitchik wrote three major monographs in the early to mid 1940s, two of which were published only decades later: "Ish ha-Halakhah" (in Hebrew), *Talpiot* 1: nos. 3–4 (1944): 651–735, translated into English by Lawrence Kaplan, *Halakhic Man* (Philadelphia: Jewish Publication Society of America, 1983); "U-Bikashtem mi-Sham," in *Ish ha-Halakhah: Galui ve-Nistar* (Jerusalem: World Zionist Organization, 1979); *The Halakhic Mind: An Essay on Jewish Tradition and Modern Thought* (New York: Seth Press, 1986). See also "The Lonely Man of Faith," *Tradition* 7, no. 2 (1965): 5–67; "The Community," *Tradition* 17, no. 2 (1978): 7–24; "Majesty and Humility," ibid., 25–37; "Catharsis," ibid., 38–54. In addition, a growing number of Soloveitchik's oral presentations and *shi'urim* (lessons) are being published, based on notes and recordings.

The World of Rabbi Joseph B. Soloveitchik

Any introduction to Soloveitchik's thought requires an understanding of his philosophical framework. His philosophical approach is phenomenological, in the tradition of Husserl. Given his concern with phenomenology, Soloveitchik is not interested in addressing many of the problems that concerned medieval Jewish philosophers. Soloveitchik is a modern philosopher working in a post-Humeian, post-Kantian world. Since such questions as the nature of God, the nature of the world, and the nature of man in any absolute sense have all been shown to be futile and beyond the scope of rational human inquiry, Husserlian phenomenology focuses instead on human experience. Thus Soloveitchik's works describe the religious experience, rather than offer classic philosophical analyses. All that we know of the world is our experience of it. The role of philosophy is then to provide an in-depth description of human experience. Soloveitchik also seeks typological descriptions, like those proposed by the German philosophers Edouard Spranger and Wilhelm Dilthey, in which idealized human types are defined and described. For these thinkers, as for Soloveitchik, the described types are abstractions, whereas real people are always composites of various types. Finally, dialectical reasoning features prominently in Soloveitchik's writings; in this he follows Hegel, Kierkegaard, and Rudolph Otto.

In *Halakhic Man* (1944), Soloveitchik presents a complex phenomenological description of the rich inner and outer life as lived by halakhically observant Jews. It is a unique work in the annals of rabbinic literature, best described by Lawrence Kaplan in his introduction to the English translation:

> The wide-ranging nature of *Halakhic Man* calls forth, nay requires, the full deployment of R. Soloveitchik's vast erudition. From a discussion of a biblical or talmudic passage he may smoothly and almost imperceptibly move to an analysis of modern scientific method, then turn to an exposition of Aristotelian or Maimonidean philosophy, buttressed by appropriate modern historical scholarship, follow up with reference to modern secular and religious phenomenology and existentialism, and cap the discussion with an acute resolution of a knotty halakhic issue, citing appropriate medieval and modern rabbinic scholarship. Heidegger, Kant, Hermann Cohen, Scheler, Barth, Cassirer, Einstein, Planck and Niebuhr rub elbows with R. Hayyim Volozhin, R. Isaac of Karlin, R. Joseph Babad, R. Lipele of Mir, R. Isaac Blaser, R. Naphtali Tzvi Yehudah Berlin, R. Hayyim Heller, and the members of R. Soloveitchik's own distinguished rabbinic family: his grandfathers, R. Hayyim Soloveitchik and R. Elijah of Pruzhan; his uncles, R. Menachem Krakowski and R. Meir Berlin (Bar-Ilan); and his father R. Moses Soloveitchik. Above this colorful and varied throng hovers Maimonides, both the Maimonides (or, perhaps better,

> the Rambam) of the *Mishneh Torah*, and the Maimonides of the *Guide of the Perplexed*, together with their "armor-bearers," medieval and modern, aharonim and practitioners of Jüdische Wissenschaft.[2]

The work includes terse analytic discourse, as well as "rapturous poetic outbursts,"[3] as it seeks to define and differentiate the ontological outlooks of three ideal types: cognitive or scientific man,[4] religious man,[5] and halakhic man.

Soloveitchik characterizes the cognitive approach of scientific man as follows:

> When scientific man observes and scrutinizes the great and exalted cosmos, it is with the intent of understanding and comprehending its features; scientific man's desire is to uncover the secret of the world and to unravel the problems of existence. When theoretical and scientific man peers into the cosmos, he is filled with one exceedingly powerful yearning, which is to search for clarity and understanding, for solutions and resolutions. Scientific man aims to solve the problems of cognition vis-à-vis reality and longs to disperse the cloud of mystery which hangs darkly over the order of phenomena and events. . . . He desires to establish fixed principles, to create laws and judgments, to negate the unforeseen and the incomprehensible, to understand the wondrous and the sudden in existence.[6]

Furthermore, "Cognition, for him, consists in discovering the secret, solving the riddle, hidden, buried deep in reality, precisely through the cognition of the scientific order and pattern in the world. In a word, the act of scientific man is one of revelation and disclosure."[7]

2. Soloveitchik, *Halakhic Man*, p. viii. "R." is an abbreviation for "Rabbi."
3. Ibid., ix.
4. In his excellent translation Kaplan translates Soloveitchik's "*ish ha-da'at*" as cognitive man. While it is true that Soloveitchik has a broad understanding of the activities of cognitive man, the epitome of cognitive man for Soloveitchik would be a twentieth-century mathematical physicist, such as Einstein, Schrödinger, or Bohr. Since most of the examples of cognitive man's ontological approach come from science, I lean toward translating *ish ha-da'at* as scientific man, perhaps displaying my professional bias.
5. Kaplan uses the term *homo religiosus* for Soloveitchik's "*ish ha-dat*." The parallel translation of "*ish ha-da'at*" (discussed in my previous footnote) would be *homo scientificus*. While it is certainly in line with Soloveitchik's general style, both in Hebrew and English, to intersperse technical terminology, such as *homo religiosus* and *homo scientificus*, in his writing, parallel usage would demand *homo halakhicus* for halakhic man. I prefer to speak simply of religious man, scientific man, and halakhic man. Although all of my quotes from *Ish ha-Halakhah* are based on Kaplan's translation, I have taken the liberty of making some minor modifications.
6. Soloveitchik, *Halakhic Man*, 6.
7. Ibid., 7.

It is important to note that, for Soloveitchik, scientific man is far from the stereotypical coldhearted, atheistic man of science, as is often depicted by religious writers. Instead, he points out that throughout most of the history of Western science, the great thinkers, like Newton and Galileo, were seeking to understand God by developing a profound and deep understanding of the orderly working of his universe. What characterizes scientific man is not his rejection of God; rather, he seeks God by appreciating God's immanence in the world, as manifested in the laws of nature.

In what way, then, does religious man differ from scientific man? According to Soloveitchik:

> When he [religious man] confronts God's world, when he gazes at the myriads of events and phenomena occurring in the cosmos, he does not desire to transform the secrets embedded in creation into simple equations that a mere tyro is capable of grasping. On the contrary, religious man is intrigued by the mystery of existence—the *mysterium tremendum*—and wants to emphasize that mystery. He gazes at that which is obscure without the intent of explaining it and inquires into that which is concealed without the intent of receiving the reward of clear understanding. The dynamic relationship that exists between the subject-knower and the object-known expresses itself, for religious man, not in the desire and ability of the subject to comprehend the object but, on the contrary, in accepting the fascinating, eternal mystery that envelops the object.[8]

Soloveitchik's types are not two-dimensional cardboard figures. Instead, they are active agents pursuing different types of intellectual enquiry:

> This is not to say that religious man prefers the chaos and the void to the structured cosmos or that he would choose to undo the act of creation and introduce confusion into reality. Heaven forbid! Religious man, like cognitive man, seeks the lawful and the ordered, the fixed and the necessary. But for the former, unlike the latter, the revelation of the law and the comprehension of the order and interconnectedness of existence only intensifies and deepens the question and the problem. For while cognitive man discharges his obligation by establishing the reign of a causal structure of lawfulness in nature, religious man is not satisfied with the perfection of the world under the dominion of the law. For to him the concept of lawfulness is in itself the deepest of mysteries. Cognition, according to the world view of the man of God, consists in the discovery of the wondrous and miraculous quality of the very laws of nature themselves.[9]

Thus religious man is enthralled by the mystery of the universe, no less than is cognitive man. However, religious man is not satisfied with the scientific,

8. Ibid.
9. Ibid.

quantitative description of the cosmos. The more he learns about the world, the more he aspires to a higher world beyond that perceived by the senses and beyond that captured by science. This leads religious man to seek transcendence, a desire to escape the mundane constraints of the here-and-now and to find refuge in the company of the Other, who is beyond the physical world. In following his rapturous desire to find God-who-is-not-in-the-world, religious man is constantly seeking to flee the superficial world of the physicist in order to dwell in the more "real" world that transcends our senses.[10] In fleeing the constraints of this world, religious man becomes increasingly more subjective in interpreting his experiences.

What of halakhic man? Where does he fit into this dichotomy? Soloveitchik proclaims that halakhic man shares many of the attributes of scientific man. On the other hand, he also seeks out the God who remains hidden in nature. Halakhic man is said to reflect "two opposing selves, two disparate images."[11] For Soloveitchik, halakhah is solely concerned with this world, the same world evident to the senses as that interrogated by the scientist. Halakhic man has no interest in escaping from the world; indeed, the whole point of halakhah is to learn how to act in and upon the world. Halakhah is a way of imposing order on a seemingly chaotic world. Just as the scientist tries to discover the natural laws that explain the apparently frenetic and haphazard events of the world, the halakhist seeks the moral laws that help explain the frenetic and haphazard events of the seemingly chaotic moral world. Although the methodologies of science and of talmudic study are different, Soloveitchik draws extensive parallels between the scientist and the halakhist. The great strength of halakhic Judaism for Soloveitchik is that the halakhah is an objective manifestation of religious experience, just as the physical sciences draw strength from their relative objectivity when compared with the human or social sciences. It is clear that Soloveitchik aligns the halakhist with the rational world of the scientist, rather than with the occasionally irrational approach of religious man. For example he writes:

> It is preferable that religion should ally itself with the forces of clear, logical cognition, as uniquely exemplified in the scientific method, even though at times the two may clash with one another, rather than pledge its troth to beclouded, mysterious ideologies that grope in the dark corners of existence, unaided by the shining light of objective knowledge, and believe that they have penetrated to the secret core of the world.[12]

10. Cf. Leo Baeck's essay "Romantic Religion" in Baeck, *Judaism and Christianity: Essays*, trans. Walter Kaufmann (New York: Atheneum, 1981), 189–292.
11. Soloveitchik, *Halakhic Man*, 3.
12. Ibid., 141n4.

Unlike the scientist, halakhic man is not satisfied with the world as it is, but rather than fleeing the world to become transcendent, in the manner of religious man, halakhic man endeavors to bring the transcendent down to this world. The performance of the halakhic act is a way of anchoring the transcendent in the immanent. Rather than creating disharmony, the constant dialectical tension that halakhic man encounters in experiencing both the immanent and the transcendent is the source of dynamic creativity that characterizes the halakhic life for Soloveitchik: "Out of the contradictions and antinomies there emerges a radiant, holy personality whose soul has been purified in the furnace of struggle and opposition and redeemed in the fires of the torments of spiritual disharmony. . . . The deep split of the soul prior to its being united may, at times, raise a man to a rank of perfection."[13] Soloveitchik, following Maimonides, asserts that humans can never directly apprehend God, but only God's attributes, such as mercy, compassion, and forgiveness. Man is commanded to emulate God's attributes (*hitdamnut le-el*, imitatio dei) and the halakhic system provides the roadmap to achieve this:

> When man, the crowning glory of the cosmos, approaches the world, he finds his task at hand—the task of creation. He must stand on guard over the pure, clear existence, repair the defects in the cosmos, and replenish the "privation" in being. Man, the creature, is commanded to become a partner with the Creator in the renewal of the cosmos: complete and ultimate creation—this is the deepest desire of the Jewish people.[14]

What is so unusual about Soloveitchik is his stress on the importance of creativity to halakhic life. He points out that our first introduction to God in the beginning of Genesis is as the Creator of the universe. Imitatio dei therefore requires man to strive for creativity. In the last part of *Halakhic Man*, Soloveitchik discusses at length the phenomenology of *teshuvah* (repentance) and develops a theory of *teshuvah* as the ultimate creative act, wherein man has a chance to (re)create himself.[15] Soloveitchik concludes:

> Halakhic man, whose voluntaristic nature we have established earlier, is, indeed, a free man. He creates an ideal world, renews his own being and transforms himself into a man of God, dreams about the complete realization of the Halakhah

13. Ibid., 4.
14. Ibid., 105.
15. Ibid., pt. 2. See also Pinchas H. Peli, *On Repentance: In the Thought and Oral Discourses of Rabbi Joseph D. Soloveitchik* (Jerusalem: Oroth, 1980). The concept of *teshuvah* is extremely important in Soloveitchik's thought. Each year, during the week between Rosh Hashanah and Yom Kippur, Soloveitchik delivered a lecture exploring from many different dimensions the phenomenon of *teshuvah*. These lectures lasted three to five hours and attracted thousands of listeners.

in the very core of the world, and looks forward to the kingdom of God "contracting" itself and appearing in the midst of concrete and empirical reality.[16]

This rejection of otherworldliness as the valid domain of Jewish thought, and the emphasis instead on the necessity of a deep and profound understanding of the scientific, technological, cultural, and aesthetic aspects of the world, is necessary for developing the requisite religious sensibility. This theme is repeated throughout virtually all of Soloveitchik's later writings.[17]

Rabbi Kook's Worldview

As an author, Soloveitchik was a perfectionist, devoting much time to editing and reediting his manuscripts and never feeling that they were ready for publication. Kook, on the other hand, was a compulsive writer and was so busy composing new works, writing his diary, and corresponding with others that he never had time to go back and edit what he had already written: "My thoughts reach wider than the sea; I cannot express them in prose. In spite of myself, I am compelled to be a poet—a free poet unfettered by meter and rhyme. I flee from conventional prose because it is ponderous and limiting, nor can I tolerate other restrictions that may even outweigh the fetters of prose from which I flee."[18] At times terse, always poetic, his writings are filled with words and allusions to the Bible and the entire gamut of rabbinic literature, making them difficult to read and virtually impossible to capture in English translation.

Kook's primary intellectual debt was to Hegelian dialectics. The history of the cosmos, as well as the history of civilization, is characterized by continual movement and progress. This progress is best understood as the confrontation between opposing values and ideas—between thesis and antithesis. The clash of ideas lasts for a finite period of time, before the world progresses by the creation of a new all-encompassing model that represents the synthesis of the earlier ideas.[19] Kook's thinking revolves around such antitheses as good and evil,

16. Ibid., 137.

17. See particularly Soloveitchik, "U-Bikashtem mi-Sham"; Soloveitchik, "Lonely Man of Faith"; Soloveitchik, "Majesty and Humility"; Soloveitchik, "Sacred and Profane: Kodesh and Chol in World Perspective," in *Gesher* 3, no. 1 (1966): 5–29.

18. *Perakim be-Mishnato ha-Iyyunit shel ha-Rav Kook*, ed. Y. Hadari and Z. Zinger (Jerusalem: Amanah, 1961), 55.

19. While Kook's dialectic is clearly based on his understanding of Hegel, this pattern of thinking is also found in kabbalah, and it is undoubtedly due to this conjunction that Kook adapted this mode of analysis in so many of his works. Unlike Soloveitchik, Kook's knowledge of philosophy was almost certainly drawn from secondary sources.

matter and spirit, chance and necessity, light and dark, monotheism and pantheism, belief and doubt, deed and thought, variation and unity, and, especially, the sacred and the profane. From his Hegelian standpoint, Kook is therefore able to see the positive aspects of those elements that are generally regarded as an anathema by his religious peers. For example, Kook considered that good and evil are both necessary, because their conflict will push the world to a higher level, where it will experience a state that is beyond good and evil. A similar process occurs with all the other fundamentally conflicting pairs of concepts. At its profoundest the Torah expresses the ultimate synthesis of all opposites, because it represents the thoughts of God, who alone is the true *coincidentum oppositorum.*

It is easy to see why evolutionary thinking was so acceptable to Kook, who proclaimed that "evolutionary theory, which is now achieving such worldwide acclaim, coincides with the lofty doctrines of kabbalah *more than any other philosophical doctrine.*"[20] For Kook, the world is constantly evolving as the divine plan unfolds. The world began as *tohu va-vohu* (unformed and void), and since the Creation it has been evolving in a basically progressive direction—sometimes moving as a violent wave and sometimes as a trickle, sometimes the forward motion is apparent, while at other times it zigzags. But ultimately the world will reach a state of perfection that the Torah describes as the messianic era. Kook viewed human history as mirroring this constant progress and ceaseless growth. This process can appear unsettling, but it represents God's profound blessing, which outweighs any need for stability.[21] Both the material world and the spiritual world undergo perpetual movement, which, while not always in consonance with each other, ultimately denotes an ascent, either to higher levels of spirituality or to higher, more complex physical forms.[22]

For Kook, the growth and progress achieved in Darwinian evolution and cosmology—being the sciences that document and record biological and cosmic history, respectively—cohere well with his progressivist metaphysics. "Scientific progress signifies movement toward the transcendental divine radiance and supernal vision, wherein the particular reflects the integrated divine scheme and foundation of the cosmos."[23] Kook deflects the notion that there is any inherent conflict between evolution and religion by shifting

20. Abraham Isaac ha-Kohen Kook, *Orot ha-Kodesh*, 10 vols. (Jerusalem: Mossad ha-Rav Kook, 1985–), 2:557. Emphasis added.

21. Ibid., 1:50.

22. Ibid., 2:518.

23. Ibid., 2:532.

the focus away from the usual issues of contention, such as the deep history of the universe or the facts of evolutionary change. For Kook, religious thinking—unlike the empirical sciences—is fundamentally concerned with teleology: the primary questions that religion seeks to answer are the teleological ones, whereas the primary questions of empiric science are the material and causal ones. Therefore, in the "teleological perception of the universe, the persistent transformation of the phenomenal is judged in terms of design rather than impulse (or chance)."[24] Kook saw evolution as confirming the view that each generation improves on its predecessor in some crucial respect and that progress is cumulative, views commonly held by many evolutionists of the time.[25] Thus for Kook the study of evolution possesses positive religious value; an evolution marked by constant progress provides solid grounds for optimism. Why indeed should humankind despair in the face of perpetual improvement? In conclusion, it is important to note not only that evolutionary motifs appear in Kook's writings, but also that evolution is integral, fundamental, and indispensable to his theology.[26]

Interpreting Genesis: Rashi and Naḥmanides

The fact that neither Soloveitchik nor Kook considered it necessary to refute the claims of evolution or even to offer an extensive Jewish apologetic for evolution may come as a surprise.[27] But we should remember that the writings of Kook and Soloveitchik largely predate the evolutionary synthesis of the 1940s and '50s, with its strongly materialist underpinnings, and that the more teleological and progressivist interpretations of evolution popular earlier in the century could often be more easily incorporated into a religious worldview. More important for appreciating Kook's and Soloveitchik's receptivity to the biological theory of the evolution of species is the extent to which their responses were grounded in traditional Jewish rabbinic sources. I will now present and explicate just a few of the traditional Jewish sources as they would have been understood by Kook, Soloveitchik, and indeed anyone steeped in rabbinic literature.[28]

24. Ibid., 2:557

25. Like Henri Bergson and Pierre Teilhard de Chardin, Kook adopted an optimistic, progressive view of evolution.

26. For an excellent and more complete discussion of Kook's understanding and use of evolution see Shai Cherry, "Three Twentieth-Century Jewish Responses to Evolutionary Theory," *Aleph: Historical Studies in Science and Judaism* 3 (2003): 247–90.

27. See Abraham Isaac ha-Kohen Kook, *Igrot ha-Re'iyah*, 3 vols. (Jerusalem: Mossad ha-Rav Kook, [1961–65]), no. 91, 2:722–25.

28. I am not presenting this material in an academic/critical way, but rather precisely as these sources would be used and understood in any yeshiva. It is, if you like, a mini Torah *shi'ur* (class).

The Mishnah, the authoritative third-century CE legal text edited by Rabbi Judah ha-Nasi, states: "It is forbidden to explicate the chapter dealing with sexual misconduct to three students at one time; nor the chapter on the story of creation to two students; nor [Ezekiel's] chariot to a single student unless he is wise and can understand for himself."[29] The Mishnah was written at a time when a large portion of Torah was taught orally from teacher to student. There were two facets of this *Torah she-be-al-peh* (Oral Law) that were considered to be esoteric teachings and could therefore be imparted only to students who have achieved the highest level of education and are educationally and intellectually prepared to receive it. The first of these is the *ma'aseh bereshit* and refers to the explication of the beginning of *Bereshit*—the Hebrew title of the book of Genesis—which narrates the Creation. Maimonides identified *ma'aseh bereshit* with cosmology.[30] The second and deeper level of esoteric learning is called *ma'aseh merkavah* and refers to the vision of the divine chariot seen by the prophet Ezekiel; for Maimonides this form of learning is the source of theosophy—true wisdom of God. For the rabbis of the Mishnah, the secret of Creation lay concealed behind the simple meaning of the words.

Let us examine the beginning of this troublesome text and see some of the problems identified by two medieval commentators, Rashi (Rabbi Shlomo ben Yitzḥak, 1040–1105) and Naḥmanides (1194–1270), whose commentaries appear in most editions of the Torah used in yeshivot.

Genesis 1:1–5:

1. In the beginning God created the heavens and the earth.
2. And the earth was without form and void, darkness was on the deep and the spirit of God hovered on the face of the waters.
3. God said "Let there be light," and there was light.
4. God saw the light was good; God divided between the light and the dark.
5. God called the light day and the darkness night; and there was evening and there was morning, one day.

These familiar words are the opening of the Torah. On the surface they seem simple enough. The textual problems facing the medieval expositors were the meanings of individual words and the temporal sequence of Creation. From these verses, what should we infer that God created first? When I pose this question to my freshman biology class at Yeshiva University, I usually have a 50–50 split, with half of the class responding "light" (verse 3) and the other half "heaven and earth" (verse 1). I inform them, in the finest

29. Mishnah Ḥagigah 2:1.
30. Moses Maimonides, *Hilkhot Yesodei ha-Torah* 1:6-10.

yeshiva tradition, that they are both right or, to be more precise, that neither of them is wrong. This difference in interpretation was exactly the argument between Rashi and Naḥmanides. First, let us hear Rashi's view:

> This verse [Genesis 1:1] cries out to be interpreted. . . . If you want to explain its literal meaning, this is the way to understand it. In the beginning of the Creation of heaven and earth, when the heaven and the earth were yet unformed and void. . . . God said "Let there be light." And it does not come to tell us that heaven and earth preceded light, for if that is what it wanted to convey it would have said "At first (*be-rishonah*) God created the heavens and the earth." The word *reshit* [is in the construct form and] always connotes in the beginning *of* something [and Rashi gives other examples from the Bible] . . . . For if you want to say that that heaven and earth were created first, . . . then ask yourself, "Where did the water [verse 2] come from?" Clearly these first two verses do not come to tell us anything about an order of which came first or which came later at all.[31]

Based on his linguistic analysis, Rashi is convinced that light was the first creation, and the first two verses are to be read as an introductory lead-in.

Naḥmanides, however, takes issue with Rashi in the extreme. His objections are both linguistic and substantive. The linguistic critique is too complex to be discussed here, but it leads to a very different reading of the text.

> And now hear the clear and correct meaning of these verses. The Holy One, Blessed is He, created all things from absolute nothingness. In the Hebrew language the only verb that can describe this ex nihilo creation is *bara*. Not everything that is now seen below or above the heavenly sphere was created [directly] from this absolute nothingness. Rather, God drew out of this complete and absolute nothingness a very thin primordial material, without any real substance to it, but that had the ability [or power] to become actualized and to attain shape [or substance], and to move from the potential to the actual. And this first primordial matter was what the Greek philosophers called *hylum*. After this creative act God did not "create" [in the sense of *bara*] anything else. The rest of creation [described in this chapter] uses terms such as "made" or "formed," because all else was drawing out and giving form to the earlier primordial material.
>
> If so, then the correct meaning of this text is: At first God created *shamayim* [heaven], for it was created from nothingness, and *aretz* [earth], which was also created from nothingness, and this "aretz" included the four basic elements. . . . In this creation, which began as a small, fine point, was included all that will [eventually] become the heavens and the earth.[32]

31. Rashi on Genesis 1:1.

32. Naḥmanides on Genesis 1:1.

For Naḥmanides, the first two verses of Genesis describe an actual part of the Creation process, in which God creates the primordial material, which develops into the material world as we see it. This "thin and unformed" primordial material, which is described as *tohu va-vohu*, was created as a single point and represents God's only absolute creative act. During the course of the rest of the Creation process, things are "made" or "formed" by God, but not created. So the world as we know it arises by a process of development or unfolding from earlier forms, a view that possesses certain affinities with the theory of the evolution of species.

In addition to his philological disagreement with Rashi, Naḥmanides has other reasons for seeking evidence of earlier forms of creation at the beginning of the Torah. Naḥmanides was working within a long rabbinic tradition with roots in the *Sefer ha-Bahir*,[33] attributed to a first-century Mishnaic sage, Rabbi Neḥuniah ben ha-Kanah. In this earliest of kabbalistic works, the cosmology of the universe is compared to the agricultural cycles that are halakhically mandated for the Land of Israel, known as the *shmittah* (sabbatical) cycle. In this cycle, the land may be worked for six years, but in the seventh year the land is rested and no planting, watering, fertilizing, or other forms of agricultural work can be done. After seven of these cycles have occurred, the fiftieth year is a Jubilee year, in which all lands are returned to their original owners, all debts are canceled, and bondsmen are freed. The cosmos also goes through its own universal *shmittah* cycles, but each year of the earthly *shmittah* cycle is represented by a thousand years in the divine or universal *shmittah*.[34] According to this tradition there will be seven such universal cycles before the world achieves its highest level of perfection, the universal Jubilee, the messianic era. The history of the world from Adam and Eve to the present represents but a single cycle. According to Naḥmanides, the existence of these earlier cycles is hinted at, but concealed, in the first two verses of the Torah. Therefore, in his dispute with Rashi, he would not concede that the Creation described in Genesis was the first and only creation.

Furthermore, while Rashi sees the unfolding of Creation taking place in time, Naḥmanides disagrees.

33. While the *Sefer ha-Bahir* is an esoteric work, the concept of universal *shmittot* (discussed later in the paragraph) was incorporated into the Babylonian Talmud (Sanhedrin 97a) and the Midrash Rabbah and therefore became part of normative rabbinic thought.

34. According to at least one school of thought these universal "years" are made up of universal or divine "days," each one of which is a thousand years long. If so, then the seven universal *shmittah* cycles would last in the range of fifteen billion years. See Aryeh Kaplan, *Immortality, Resurrection, and the Age of the Universe: A Kabbalistic View* (Hoboken, NJ: Ktav, 1993), 1–15.

> In its in-depth meaning the term "day" refers [not to a period of time but] to the divine emanations that are called sefirot. For each creative act of God, in which more of the divine is revealed in the world, is called a "day". . . . And the explanation of all this is very deep and profound and our ability to explicate it is less than a drop in the entire ocean.[35]

Therefore for Naḥmanides, the sequence of the six days of creation is not a chronological or temporal sequence, but describes different levels of divine revelation that kabbalists called the sefirot (emanations).

I want to emphasize that while Naḥmanides' commentary on the Torah is second in importance and popularity to that of Rashi, anyone advancing beyond the most rudimentary traditional Jewish education would be exposed to Naḥmanides. Furthermore, while his commentary on Genesis appears at first glance to be the more radical of the two, traditional Jews view Naḥmanides as a far more conservative commentator than Rashi, because he always tries to explain texts in conformity with traditional rabbinic dicta. In presenting this very nonliteral translation of Genesis, Naḥmanides is seeking to remain within the common rabbinic tradition, as found in the standard midrashic sources (rabbinic comments and interpretations of passages in the Torah) of which I will mention just two, both of which are from the Midrash Bereshit Rabbah, dating from about the fifth century CE.

> Rabbi Judah bar Simon says: It does not say "there was evening" but "*and* there was evening." This teaches us that there were time sequences prior to the one being described. Rabbi Abbahu says: From this we learn that God created worlds and destroyed them, created and destroyed them, saying "This one pleases me, this one does not please me."[36]

> "And there was evening, and there was morning, the sixth day [not *a* sixth day]": Rabbi Simon bar Marta said until now [the end of the sixth day] the counting was according to a universal time, from here onward there is a different time scale.[37]

In the light of such texts as these, the scientific findings of the nineteenth and twentieth centuries that pointed to a long and slowly developing history of

35. Naḥmanides on Genesis 1:3.

36. Midrash Bereshit Rabbah 3:7. The theological implications of this particular midrash are quite radical and provocative. An all-knowing, all-powerful God creates a world that ends up not pleasing him! This implies a significant degree of autonomy and/or nondeterminism in the created world.

37. Midrash Bereshit Rabbah 9:14.

both the cosmos and the biological world should not present a problem for Jewish thinkers.

Yisroel Lifschitz: Genesis and the Development of Life

Given this background of rabbinic thinking it is not surprising that Kook and Soloveitchik were not the first or only traditional rabbinic scholars to embrace an evolutionary perspective. A mid-nineteenth-century example is Rabbi Yisroel Lifschitz, who engaged the latest scientific findings about the development of life on earth, as revealed by geology and paleontology.[38] A rabbinic scholar of tremendous influence, Lifschitz was rabbi in Danzig (Gdansk) from 1837 to 1860. His commentary on the Mishnah, the *Tif'eret Yisrael* (Beauty of Israel) has become one of the three standard commentaries that is routinely published with each new edition of the Mishnah. Lifschitz wrote a remarkable monograph first published in 1840[39] called *Derush Or ha-Ḥayyim* (Sermon on the Light of Life). Originally composed as a sermon for his congregation and delivered during the intermediate Sabbath of Passover, the work deals in a broad and comprehensive way with Jewish eschatology.[40] Lifschitz exposed his congregation to the various terms used by the rabbis to describe the end of days. He differentiated between such concepts as the messianic era (*yemot ha-mashiaḥ*), the world to come (*olam ha-ba*), and the resurrection of the dead (*teḥiyyat ha-metim*) and surveyed the different rabbinic opinions of what these terms mean. In the course of the sermon, Lifschitz claimed that just as the Torah and rabbinic tradition have specific teachings about the future of the universe, we also have been given a glimpse of the history of the universe. He described for his listeners the traditional understanding of *ma'aseh bereshit*, as including earlier creations before our own and explained the idea of the universal *shmittot*, with God creating and destroying prior worlds.

He then stated: "My dear brethren, see on what a firm basis the teachings of the Torah stand, for this knowledge [of ancient worlds] that has been passed down orally in our tradition for hundreds of years, is only now being

38. Lifschitz was the first to develop this approach, which had been suggested in the late eighteenth century by Pinḥas Elijah Hurwitz; see Ira Robinson, "Kabbala and Science in *Sefer ha-Berit*: A Modernization Strategy for Orthodox Jews," *Modern Judaism* 9 (1989): 275–88.
39. Since evolution did not become widely accepted until after Darwin published his *Origin of Species* in 1859, Lifschitz's sermon cannot be seen as a Jewish apologia, attempting to make Judaism conform to a popular trend. Obviously it also does not discuss the idea of natural selection introduced in the *Origin*.
40. The underlying theme of Passover is, of course, rebirth and renewal.

discovered by the scientists of our time." He went on to describe in some detail the early discoveries of geological stratification, as well as the remarkable discovery of the giant dinosaurs and mammoths, whose petrified remains he urged his congregants to see in the Zoological Museum in St. Petersburg. He pointed out that some of the dinosaurs were herbivores, like Apatosaurus, and some were carnivores, like Tyrannosaurus Rex. He also argued that both geological discoveries (for example, that huge rocks moved from place to place) and biological evidence (such as woolly mammoths being found in Siberia) prove that tremendous upheavals and cataclysms had occurred during the earth's history. Then he indicated that, although the earlier prehistoric animals were in general larger than those in more recent strata, it is possible to discern a progressive development in their bodily forms in the more recent strata. For Lifschitz, science confirmed the truth of Torah and of rabbinic tradition. Lifschitz was not even surprised by the discovery of earlier forms of human being and referred to these as pre-Adamite humans. He offered his own brilliant reading of the first verses of Genesis, and stated, in a similar fashion to the Naḥmanides, that this information is included in the Torah as an esoteric message. The main purpose of Torah is not to teach cosmology, although it obviously cannot contradict accurate cosmology, but rather to show how we must behave in order to be God's partners in the universe. Since cosmic history is not terribly germane to human behavior, the Torah rapidly focuses on what Lifschitz called *adam ha-rishon ha-akhshavi*, or modern man.[41]

The history of the *Derush Or ha-Ḥayyim* is a happy one. It has not been consigned to dusty library shelves; rather, the editors of the Yakhin u-Bo'az Mishnah, which has long been the standard rabbinic edition, chose to include this sermon, in its entirety, at the end of the first volume of Nezikin (Damages). Therefore, the sermon has found its way not only into every Jewish library, but also into virtually every Orthodox yeshiva and school in the world. Alas, that apparently so few have had the curiosity to actually read it!

Concluding Personal Remarks

I am by profession a biologist, trained in microbiology and immunology, and have been involved in cancer research for my entire professional career. My avocation for science showed up early and the first Passover afikoman present that I remember asking for as a child was a chemistry set. The next year

41. Kook explicitly uses this idea in *Igrot ha-Re'iyah*, no. 91, 2:722–25.

it was a microscope. I am also an Orthodox Jew who grew up in a house that was rich in traditional observance of the Sabbath, Jewish holy days, and all the other mitzvot (religious precepts). My father's library was filled with the Torah and a wealth of classic halakhic and aggadic (nonlegal) texts, as well as more modern works on Jewish history, Hebrew language, and Jewish philosophy. I learned the Hebrew alphabet at the same time that I learned Dick and Jane. The library also had a complete gamut of classics of Western literature. The result was that I lived and imbibed Torah u-madda (Jewish and general studies) long before I heard that term used. For me, evolution was never particularly problematic. Reading *Inherit the Wind* as an adolescent, I was clearly on the side of the young biology teacher. The William Jennings Bryan character was a Christian fundamentalist whose simplistic and literal reading of Genesis bore no resemblance to the Genesis that I had learned with Rashi and Naḥmanides. Doesn't the Mishnah state that the first chapter of Genesis describes the mystery of God's creative act, the secret of *ma'aseh bereshit*? If it were intended to be understood literally, then what mystery remains?

Later in my life, the "arguments" of anti-evolutionists, creation scientists, and advocates of Intelligent Design had no resonance for me. For this reason I maintain that that there is no Jewish "problem" with the science of evolution. As I believe has been amply documented in this paper, Abraham Isaac Kook and Joseph B. Soloveitchik, the two most important and influential Orthodox Jewish thinkers of the twentieth century, who based their analyses on fairly traditional readings of classic Jewish texts, not only dismissed the notion of any conflict between modern science and Torah, but actually found contemporary scientific notions of evolution and cosmology to be harmonious with classic rabbinic thought.

10

The Order of Creation and the Emerging God: Evolution and Divine Action in the Natural World

Lawrence Troster

On my way to a local Jewish Community Center, I pass a church that displays an inspirational message on its sign. One day it read: "Because God is in control we have nothing to fear." I was struck by this assertion, which contains one of the most contentious issues of modern theology. How do we know that God is in control of the world? And if so, what kind of control are we talking about? Do we mean general concern for the world, a grand cosmic plan, ongoing creation, or individual care for each human being? The sign does not say.

This issue is usually described as the doctrine of providence or divine action in the world. Gershom Scholem, in an article surveying modern Jewish theology, wrote:

> What I have in mind are the attributes of omnipotence and providence, which are allegedly evident in God's acts and which are in sharpest contradiction with the human freedom of moral decision. This freedom of decision, however, is the basis of the moral world of Judaism, which stands and falls with it, today as it did 3,000 years ago. Nowadays, the least plausible of all "dogmatic" assertions of Jewish theology is the thesis of the providence of God, who in His infinite, all-embracing wisdom is supposed to have foreseen not only the meaning of Creation but also its development in every detail and at every stage. Even those convinced of God's existence will find it hard to come to terms with this doctrine.[1]

Why has providence become the "least plausible of all 'dogmatic' assertions of Jewish theology"? When surveying modern Jewish thought, one can see that the brute fact of the Holocaust has had the biggest impact. Richard Rubenstein wrote that to see the Holocaust as part of a divine plan for

I would like to thank my friend and teacher Rabbi Neil Gillman for his help and support.

1. Gershom Scholem, *On Jews and Judaism in Crisis* (New York: Schocken, 1978), 281.

human history is to portray Hitler and the SS as "instruments of God's will."[2] Thus, in responding to the Holocaust, recent Jewish discussions of divine power have adopted a moral perspective. While there have been many Jewish responses to the Holocaust, there seems to be a general tendency for theologians to seek to preserve the goodness of God, human free will, and moral responsibility, even if this means limiting divine power.[3]

While some Christian theologians have also tackled the moral challenge to traditional notions of divine power,[4] they have in addition addressed the impact of modern science—whether of cosmology, quantum physics, chaos theory, or evolutionary biology. In the last fifteen years, the Vatican Observatory, in partnership with the Center for Theology and the Natural Sciences in Berkeley, California, held a series of conferences on divine action and the natural sciences, resulting in the publication of a series of books. These conferences were initiated by a speech of Pope John Paul II on the three hundredth anniversary of the publication of Isaac Newton's *Philosophia Naturalis Principia Mathematica* (1687). The pope called for a serious dialogue between religion and science, involving scientists and theologians, both Catholic and Protestant.[5] There is no comparable Jewish dialogue.

2. Richard Rubenstein, *After Auschwitz: Radical Theology and Contemporary Judaism* (Indianapolis: Bobbs-Merrill, 1978), 153. Rubenstein modified his views in the second edition of *After Auschwitz* (Baltimore: Johns Hopkins University Press, 1992), 72, 174, 293–306.

3. For a good survey, see Harold Leaman, *Evil and Suffering in Jewish Philosophy* (Cambridge: Cambridge University Press, 1995). Cf. also Pinchas Peli, "In Search of Religious Language for the Holocaust," *Conservative Judaism* 32 (winter 1979): 2–24. Examples of significant theological responses to the Holocaust include Harold S. Kushner, *When Bad Things Happen to Good People* (New York: Schocken, 1981); Irving Greenberg, "Voluntary Covenant," in *CLAL Perspectives* (New York: CLAL, 1982); Harold M. Schulweis, *Evil and the Morality of God* (Cincinnati: Hebrew Union College Press, 1985); Eliezer Berkovits, *Faith after the Holocaust* (New York: Ktav, 1973); Emil Fackenheim, *God's Presence in History: Jewish Affirmations and Philosophical Reflections* (New York: Harper Torchbooks, 1972).

4. See, for example, John F. Haught, *God after Darwin: A Theology of Evolution* (Boulder, CO: Westview, 2000), 45–46; Susan Neiman, *Evil in Modern Thought: An Alternative History of Philosophy* (Princeton: Princeton University Press, 2002).

5. Robert John Russell, William R. Stoeger, and George V. Coyne, eds., *Physics, Philosophy, and Theology: A Common Quest for Understanding* (Vatican City: Vatican Observatory, 1988); Robert John Russell, Nancey Murphy, and C. J. Isham, eds., *Quantum Cosmology and the Laws of Nature: Scientific Perspectives on Divine Action* (Vatican City: Vatican Observatory; Berkeley: Center for Theology and the Natural Sciences, 1993); Robert John Russell, Nancey Murphy, and Arthur R. Peacocke, eds., *Chaos and Complexity: Scientific Perspectives on Divine Action* (Vatican City: Vatican Observatory; Berkeley: Center for Theology and the Natural Sciences, 1995); Robert John Russell, William R. Stoeger, and Francisco J. Ayala, eds., *Evolutionary and Molecular Biology: Scientific Perspectives on Divine Action* (Vatican City: Vatican Observatory; Berkeley:

The purpose of this paper is to examine the Jewish concept of providence, or divine action, in light of the modern theories of biological and cosmic evolution. While recognizing that this topic raises important issues for the much-debated question of evil, especially among post-Holocaust theologians, these large themes will not be directly addressed here. Instead, I will first summarize the concept of providence in Jewish theology and draw some comparisons with Christian theology. I will then outline the challenge of modern evolutionary thought to the concept of divine action. Then, using Ian Barbour's typology of theological models of divine action, I will focus on three thinkers: scientist Paul Davies, philosopher Hans Jonas, and theologian John F. Haught. Their contributions to this subject can, I believe, provide us with a basis for creating a Jewish theology of divine action that is compatible with evolution. This project can be seen as complementing those theologies of providence that focus on the moral challenge of the Holocaust. The implications of evolution for Jewish concepts of divine action have not received significant consideration in modern Jewish theology, and so I hope to further discussion of this concept.[6]

In formulating a response to evolution, it is also important to decide how such a response can be judged as authentically Jewish. In this I follow my teacher Neil Gillman in two ways. First, he has said that "true" theology arises from those "who share a sense of tradition that has become problematic and yet holds out the promise of renewed meaning."[7] Second, I follow his example in looking to the traditional liturgy as "the locus for the authoritative system of Jewish belief."[8] In the morning service the Shema, an evocative prayer, is surrounded by blessings in which God is defined as Creator (both initial and ongoing), Revealer, and Redeemer.[9] These are the classic Jewish theological

Center for Theology and the Natural Sciences, 1998); Robert John Russell, Nancey Murphy, Theo C. Meyering, and Michael A. Arbib, eds., *Neuroscience and the Person: Scientific Perspectives on Divine Action* (Vatican City: Vatican Observatory; Berkeley: Center for Theology and the Natural Sciences, 2000); Robert John Russell, Philip Clayton, Kirk Wegter-McNelly, and John Polkinghorne, eds., *Quantum Mechanics: Scientific Perspectives on Divine Action* (Vatican City: Vatican Observatory; Berkeley: Center for Theology and the Natural Sciences, 2002).

6. A significant exception is Arthur Green, *Seek My Face, Speak My Name: A Contemporary Jewish Theology* (Northvale, NJ: Aronson, 1992), esp. 54–55, 71–74.

7. Neil Gillman, *Sacred Fragments: Recovering Theology for the Modern Jew* (Philadelphia: Jewish Publication Society, 1990), p. xxvi.

8. Neil Gillman, *The Death of Death: Resurrection and Immortality in Jewish Thought* (Woodstock, VT: Jewish Lights, 1997), 126. Gillman here is speaking about how the concept of resurrection became canonized by being embedded in the *Amidah*, which Jews have to recite daily.

9. See for example, Reuven Hammer, *Or Hadash: A Commentary on Siddur Sim Shalom for Shabbat and Festivals* (New York: Rabbinical Assembly and United Synagogue of Conservative Judaism, 2003), pp. xxi–xxiv.

concepts in which God's providence is expressed. Thus any Jewish theology of evolution and divine action must in some way have these three elements. Finally, in looking to modern science and non-Jewish theologians, I am following what I believe to be the best methodology that Jewish theologians can adopt: to synthesize the best philosophy and science of the day with traditional Jewish concepts and sources. Throughout history, this method has produced most of the significant innovations in Jewish theology. As Maimonides wrote, "one should accept the truth from whatever source it proceeds."[10]

Traditional Jewish Theologies of Divine Action

Why is evolution such a problem for divine action? In order to answer this question, it is necessary to define what divine action meant before Darwin and then to show how neo-Darwinism has challenged this traditional definition. Classically, providence has three components: foresight, direction, and care. God foresees and governs the world, which is the object of God's care. In addition, God's providence extends to nature and to history, with God acting through the laws of nature as well as through supernatural miracles.[11]

In Jewish theology, providence was also seen as God's particular concern for the Jewish people. Providence (or in Hebrew, *hashgaḥah*, derived from Psalm 33:14: "From his dwelling-place he [intently] gazes on all the inhabitants of the earth") is a biblical concept. Rain, fertility, and victory in war were not impersonal natural or historical acts but signs of God's favor for the loyalty and righteousness of humans. There is not only the initial Creation but also an ongoing creation that provides a continuing basis for all that occurs in natural and human history (Genesis 45:5, Job 38:22–39:30, Psalms 148:8–10, Isaiah 26:12).

For Christian patristic and medieval writers, such as Augustine and Aquinas, God is the first cause of all events; all natural causes are secondary causes through which God acts. God can also intervene directly through miracles. This concept of divine action raises the double-agency problem: can a single act come from two free agents (God and human beings), each being sufficient to accomplish the event on its own? It also raises the problem of how human free will, which is finite, can exist when there is an infinite agent

10. *The Eight Chapters of Maimonides on Ethics*, ed. Joseph I. Gorfinkle (New York: AMS Press, 1966), 35–36. Examples include Philo, Maimonides, Judah Halevi, Moses Mendelssohn, Franz Rosenzweig, and Mordecai Kaplan.

11. H. P. Owen, "Providence," in *Encyclopedia of Philosophy*, ed. Paul Edwards, 8 vols. (New York: Macmillan, 1972 [1967]), 6:509–10.

(God). And there is the problem of theodicy: if God is good and omnipotent, why is there so much evil in the world?[12]

These issues were not unknown in rabbinic Judaism. In rabbinic texts divine action was considered in the light of divine knowledge, reward and punishment, human free will, and the meaning of history. Rabbinic Judaism tried to reconcile God's providence with human freedom but, according to Ephraim Urbach, the numerous references in rabbinic literature reveal no single principle that could form the basis of a systematic theology. The rabbis' agenda was rather to "activate all man's powers—both the potency inherent in the consciousness of freedom and the will to do good and that which flamed from the feeling of the nullity of man and his complete dependence on Divine Providence."[13]

Medieval Jewish discussions of providence were primarily concerned with the issues of theodicy, Islamic fatalism, and the "metaphysical problems connected with God's foreknowledge."[14] Medieval philosophers distinguished between two kinds of providence: general providence (God's care for the world or for species in general) and special providence (God's care for each individual being). For Maimonides, general providence is the norm. Special providence applies only to those human beings who were intellectually advanced, since it is through man's intellectual attainment that God becomes aware of the person. And yet Maimonides tried to preserve God's foreknowledge by insisting that human and divine knowledge are radically different. However, Gersonides, while advancing similar ideas about providence, believed that when animals perform individual acts they do so by pure chance, and that these actions are unknown to God. Gersonides also argued that a person comes under divine care as an individual only when he or she has advanced both morally and intellectually.[15]

In the modern period, according to Hillel Levine, these Jewish concepts of providence came under attack from several directions: the failure of the Sabbatean movement created a deep disappointment that left many Jews doubting the concept of divine providence; Spinoza's challenge to the concept of God's election of and providence over the Jewish people; and the rise of the scientific concept of deterministic laws of nature, which "undermined

12. Robert John Russell, Nancey Murphy, and Arthur R. Peacocke, introduction to Russell, Murphy, and Peacocke, *Chaos and Complexity*, 4.

13. Ephraim E. Urbach, *The Sages: Their Concepts and Beliefs* (Jerusalem: Magnes, 1979), 285–86.

14. Louis Jacobs, *A Jewish Theology* (New York: Behrman House, 1973), 116.

15. Ibid., 116–17; Hillel Levine, "Providence," in *Contemporary Jewish Religious Thought*, ed. Arthur A. Cohen and Paul Mendes-Flohr (New York: Free Press, 1987), 735–39; Isaac Husik, *A History of Medieval Jewish Philosophy* (New York: Atheneum, 1969), 289–94, 339–49.

the ontological plausibility of a special sphere defined by Jewish truth claims including providence."[16] As mentioned above, the idea of God's providence over the Jews and in the world in general has again become a major issue in Jewish theology after the Holocaust.

Christian theology was also influenced by the rise of modern science in the seventeenth century. This led some theologians, scientists, and philosophers to reject traditional views of divine action, especially the belief in miracles. As Newtonian mechanics developed and came to depict a causally closed universe, there was little room for God to influence individual events. Thus eighteenth-century deists insisted that God acted only at the initial Creation. The influence of Hume and Kant also undercut metaphysical speculation and tended to restrict religion to the moral sphere. And in the nineteenth century, biblical criticism further challenged the belief in miracles and other deviations from the natural order.[17] But the greatest challenge to divine action has come from Darwinism, especially in its twentieth-century synthesis with genetics (neo-Darwinism).

The Darwinian Challenge

What is it about neo-Darwinism that produces such a challenge? The Darwinian picture of the natural world denies traditional concepts of creation. Instead of being designed and created by God in its present form, the natural world, including all species, are products of a long process of natural selection. Darwinism not only provides a scientific explanation of the development of species; it also undermines the religious concept of divine providential guidance.[18]

As Catholic theologian John F. Haught has observed, Darwinism creates several critical problems for religion. First, it shows that all living beings share a common ancestry and are historically and organically interconnected. Therefore, the ontological discontinuity between humanity and the rest of creation, which was a central feature of Jewish and Christian theology, no longer exists. In addition, Darwinism challenges the traditional religious picture of the universe as a hierarchy of distinct levels of being and meaning, from the lowest levels of inanimate objects up through plants, animals, humans, and finally God, in ascending value.[19] In addition,

16. Levine, "Providence," 738–39.

17. Russell, Murphy, and Peacocke, introduction to *Chaos and Complexity*, 4–5.

18. Ian Barbour, *Nature, Human Nature, and God* (Minneapolis: Fortress, 2002), 10–12.

19. Haught, *God after Darwin*, 57–59.

Haught argues that many recent interpreters of Darwinism also see no qualitative distinctions between life and nonlife. Life arises from dead matter as a result of the laws of physics and chemistry. Thus evolutionary science can be seen to collapse and in some sense reverse the sacred hierarchy. While the traditional view was that humans are created in the image of God, the evolutionary view is that lifeless and mindless matter is the metaphysical and historical foundation of all beings, including those endowed with life and mind. Thus emergent features of the universe, like mind, are "epiphenomenal derivatives" of accidental combinations of its lowest elements.[20]

Finally, some theologians and scientists argue that the theory of natural selection asserts that the variety of life in the natural world is a product of completely random forces undirected by any intelligent agency or providential designer. The struggle for life and the survival of the fittest imply that we live in an impersonal universe.[21] These are the conclusions drawn by writers like Daniel Dennett and Richard Dawkins, who assert that Darwinism proves that there is no intelligent designer of the universe and that evolution is a completely meaningless process.[22]

Can religion therefore still claim that the universe, in the words of Haught, is "the sacramental expression of an absolute source of values?" Haught claims that theology has largely ignored the evolving cosmos partly out of a suspicion that science's view of the universe is not an objective activity but part of a modern nihilist project to desacralize the natural world for the use of technology and for profit.[23]

New Theological Models in Response

Is there an alternative worldview in which a new understanding of divine action can exist? Ian Barbour has suggested that in the twentieth century science changed from the Newtonian view of the natural world to one that is evolutionary, historical, emergent, and subject to both law and chance, structure and openness. This new view of nature is ecological, relational, interdependent, and holistic, rather than atomistic; it is organic rather than

20. Ibid, 59.
21. Ibid., 23–24.
22. Daniel C. Dennett, *Darwin's Dangerous Idea: Evolution and the Meaning of Life* (New York: Simon and Schuster, 1995); Richard Dawkins, *The Blind Watchmaker* (New York: Norton, 1986).
23. Haught, *God after Darwin*, 64–68. Cf. Seyyed Hossein Nasr, *Religion and the Order of Nature* (New York: Oxford University Press, 1996).

reductionist.[24] Within this post-Newtonian worldview, new models of divine action are possible, and Barbour elucidates four options that have been developed by various thinkers: God as designer of a self-organizing system, God as determiner of indeterminacies, God as top-down cause, and God as communicator of information.

The model of God as designer of a self-organizing system includes the notion that the overall process of evolution must also have been designed. Since such a narrow range of possibilities for life exists and because the process of evolution seems to have an inherent tendency to move toward emergent complexity, life, and consciousness, it can be inferred that the universe is designed. Design is not a detailed preexisting plan and does not include how individual species evolved; rather, design involves a general direction of growth toward complexity, life, and consciousness in which both chance and law operate. This model allows for free will, which requires both law and openness, and it understands suffering and death in the universe as intrinsic to the evolutionary process. Moreover, the experience of pain is an inescapable characteristic of greater sensitivity and awareness.

The model of God as determiner of indeterminacies relies upon quantum theory for its understanding of nature. Quantum theory specifies that indeterminacy exists in nature, which, in turn, implies that there are a range of possibilities in the universe and that the final determination might be made by God. Thus, what appears to us as chance may be the very point at which God acts. There is no energy input needed, since all the alternative potentialities in a quantum state have identical energy values. God does not intervene as a physical force, but actualizes one of the many potentialities. In order to maintain human freedom, theologian Nancey Murphy argues, God determines all quantum indeterminacies but ensures that lawlike regularities result so that there are stable structures in the physical world. Thus, scientific inquiry is possible and humans have dependable consequences for the purposes of moral choice. In human life God acts at the quantum level and also at higher levels of mental activity, but in ways that do not violate human free will.[25]

In the model of God as top-down cause, God constrains natural phenomena without violating the classical laws of nature. For example, humans

24. Ian Barbour, *Religion and Science: Historical and Contemporary Issues* (San Francisco: HarperCollins, 1977), 282–84.

25. Nancey Murphy, "Divine Action in the Natural Order: Buridan's Ass and Schroedinger's Cat," in Russell, Murphy, and Peacocke, *Chaos and Complexity*, 325–57; Nancey Murphy and George F. R. Ellis, *On the Moral Nature of the Universe: Theology, Cosmology, and Ethics* (Minneapolis: Fortress, 1996).

could be influenced by God at the highest evolutionary level, that of mental activity, without violating the lower-level laws of chemistry and physics.[26] There is a problem, however, with inanimate matter. How can God, who exists on the highest level, affect the lowest, when no intermediate levels exist? One version of this model uses the analogy of the mind-body relationship. The world is God's body and God is the world's mind. But, unlike us, God would be aware of all events in God's body, and God's actions would affect all events universally.

The model of God as communicator of information uses information theory as its basis. In electronic and biological systems, communication of information requires energy and physical input. But God, being omnipresent, would require no energy for the communication of information. In addition, since alternative potentialities are already present in the quantum world, God could convey new information without any physical input or energy expenditure. One simile is to see God acting like a choreographer who leaves much of the action to the dancers. Evolutionary history is the action of an agent who expresses intentions but does not follow an exact predetermined plan.[27]

Paul Davies

Physicist Paul Davies uses the model of God as designer of a self-organizing system, which is a modified form of the uniformitarian view of divine action.[28] While deism implies that God acts only by creating the universe and its laws at the beginning, uniformitarianism assumes that God also acts by sustaining the universe moment by moment since the universe is contingent on God's continuous creative activity.

Davies modifies the uniformitarian view of divine action in the following ways. In selecting the laws of nature God chooses "*very specific laws*" (italics in original)[29] with very remarkable properties. These laws allow not only for chance events, in the usual sense, but also for the emergence of complexity in nature, an emergence that requires these laws but goes far beyond a mere unfolding of their consequences. For Davies, complexity depends on both the deterministic outcome of the laws of nature and on a

26. Barbour, *Nature, Human Nature, and God*, 20–23, 28–30.

27. Ibid, 30–31.

28. Paul Davies, "Teleology with Teleology: Purpose through Emergent Complexity," in Russell, Stoeger, and Ayala, *Evolutionary and Molecular Biology*, 151–62. See also Davies, *The Mind of God: The Scientific Basis for a Rational World* (New York: Simon and Schuster, 1993).

29. Davies, "Teleology," 151.

"radical chance"[30] permitted by, and yet transcending, the determination of these same laws. Therefore, according to Davies, neo-Darwinism in biology, and relativity and quantum mechanics in physics, cannot explain the totality of the creativity of nature in its complexity. These laws give rise to openness in nature through natural processes having inherent powers of self-organization, which are based on these laws but not reducible to them. Therefore complexity does not require divine action to explain *particular* events. Davies, however, does see divine action in the way that these laws were set up in the first place.

Davies understands, however, that the role of chance is a "two-edged sword."[31] In most uniformitarian approaches to divine action the operation of chance would be seen as implying that God had abandoned the universe. In Davies' version, however, choice by God bestows an openness or freedom upon nature that is crucial for its creativity. Without chance, novelty could not come into existence and the world would be reduced to a preprogrammed machine.

Davies feels that the complexity of nature gives every appearance of intentional design and purpose, even though it is the result of natural processes. He calls it "teleology without teleology."[32] Divine action can be reconciled with this scientific picture of nature if viewed in the right way. He rejects the traditional interventionist model of divine action because it reduces God to an aspect of nature and is deeply antiscientific. From this perspective God becomes a "cosmic magician, who creates a flawed universe and prods it whimsically from time to time to correct the errors."[33] He also rejects the model of God as determiner of indeterminacies because the repeated action of God in this manner would violate the statistical laws of quantum physics and thereby reintroduce interventionism.

In Davies' modified uniformitarianism nature is like a game of chess, with the pieces being the physical systems and the rules being the laws of nature. The game itself is the evolution of the universe. A rich and interesting variety of scenarios is possible, but the actual outcomes are determined not only by the rules but also by the specific sequence of moves made by each player. The rules constrain and encourage certain patterns of behavior but cannot fix them in advance. The game is a mix of order and unpredictability. God influences the outcome by selecting from all the possible laws of nature

30. Ibid.
31. Ibid., 158.
32. Ibid., 159.
33. Ibid., 152.

those that are inherently statistical and that encourage rich and interesting patterns of behavior. Details of the actual evolution of the universe are left open to chance, which operates at the quantum level through chaos or through actions of the human mind.

In this version of divine action, there is no need for God to suspend, manipulate, or violate the laws of nature. There is no divine intervention, except for the miracle of existence itself. God does not exercise an authoritarian influence on the evolution of the universe. This allows both for human freedom and for inanimate systems to create true novelty in the future. God chose not to determine the universe in detail, but instead to give a vital, cocreative role to nature.

Davies rejects the idea that such a universe could have come about by chance. So long as scientists agree that there is no necessity for the universe to exist in its present state (since there could have been other possible universes), the actual universe, with its particular set of laws, must have been selected from a probably infinite set of possible laws. The contingent nature of the world raises the issues of why these particular laws were chosen and whether there is anything special about these laws. For Davies, the anthropic coincidences, the intelligibility of nature, its beauty and its harmony suggest that there is something special about them and that they are indeed the product of a loving designer. Davies rejects the idea that in a mega-universe, where there is an infinite set of universes, one universe would inevitably produce such laws. He does not accept the multiuniverse theory because an infinity of unseen universes violates Occam's razor and is unscientific—the unseen universes could not be observed and therefore could not be proved or disproved even in principle. Even if one accepts the theory of multiuniverses, it would still beg the question of how law and rationality developed from randomness. And the existence of multiple universes, each having its own laws of nature, would still need an explanation for the emergence of law itself.

Davies argues that his theory is impressively confirmed by the emergence of life, consciousness, and culture (including science and mathematics), which link the highest organizational level (mind) to lowest level (particles and fields of matter). He feels that the general trend of matter→mind→culture is written into laws of nature at a fundamental level, but that the specific details (human form, mental makeup, character of culture, etc.) depend on the accidents of evolution. Further confirmation of his model of divine action would be the existence of other life forms and cultures in the universe. If there are no other intelligent creatures in the universe, then Davies would concede that human existence is a supernatural act or a "hugely improbable but purely accidental series of events of staggering irrelevance."[34]

While Davies' model of divine action is scientifically plausible, and presents an interesting perspective on the problem of human free will, it is incomplete from a Jewish perspective. It has a very rich notion of creation, including ongoing creation, but it leaves no room for divine revelatory intervention and has no concept of redemption. For Davies, the universe has no direction other than to become increasingly complex. Nonetheless, his concept of creation could form part of a new Jewish model of divine action.

Hans Jonas[35]

Philosopher Hans Jonas created a philosophical critique of the nihilism that he felt lay at the heart of modern philosophy and then constructed a philosophical biology based on the concept of the organism and consistent with his understanding of modern science. He also created a theological model of God and divine action that not only takes science into account but also addresses the problem of evil as manifested in the Holocaust.[36]

Jonas characterized modern thought as creating new concepts of nature and of humanity that "imply the negation of fundamental tenets of the philosophical as well as religious tradition," that is, the biblical/Jewish tradition.[37] Modern, nihilist writers do not appreciate the universe as a creation, but rather as a process completely determined by mindless law. There is no longer a divine order; the world is purposeless, with no values, goals, or ends. Humanity is no longer created in the image of God. Darwinism asserts that humanity arose from impersonal forces.[38]

Jonas responded to this modern nihilism by first extending Heidegger's concept of existence to include all living organisms. Every living creature,

34. Ibid., 160.

35. Much of this section is based on my article, "Hans Jonas and the Concept of God after the Holocaust," *Conservative Judaism* 55 (2003): 16–25.

36. Jonas's major works are *The Gnostic Religion* (Boston: Beacon, 1958); *The Phenomenon of Life: Toward a Philosophical Biology* (New York: Harper and Row, 1966); *The Imperative of Responsibility: In Search of an Ethics for the Technological Age* (Chicago: University of Chicago Press, 1984); *Mortality and Morality: A Search for the Good after Auschwitz*, ed. Lawrence Vogel (Evanston, IL: Northwestern University Press, 1996). Vogel includes a bibliography of Jonas's works in English and German.

37. Hans Jonas, "Contemporary Problems in Ethics from a Jewish Perspective" in *Judaism and Environmental Ethics: A Reader*, ed. Martin Yaffe (Lanham, MD: Lexington, 2001), 250–63.

38. Ibid., 252.

from the smallest microbe to a human, shows concern for its own being. This is revealed by its relationship to the world around it in order to stave off death and nonbeing. Even in the simplest organism, there is a kind of "inward relation to their own being." For Jonas, it is metabolism—the exchange of matter with the environment that all organisms must exhibit in order to survive—that is the most basic expression of an organism's struggle for life. Every life form has what Jonas called "needful freedom," because it consistently exhibits a dynamic unity that extends beyond the sum of its parts. Yet it is also dependant on constant exchanges with its environment in order to avoid dying.[39] All the polarities that humans find within themselves—being/nonbeing, self/world, form/matter, and freedom/necessity—are to be found in the most primitive organisms.

For Jonas, every time that evolution produces a new level of complexity there is an increase in mind. This new level includes a greater freedom, as well as a greater degree of peril, and the potential for pain and suffering. For example, all life requires nutrition and the possibility of reproduction. But animal life also has the capacity for movement and desire and a level of sensitivity to its environment that most plants do not exhibit. Along with these increased capabilities, animals are endowed with the ability to feel pain, fear, and abandonment. With the advent of humanity through evolution, being now becomes reflexive and tries to understand its place in the whole of which it is part. This creates the particular human anxieties of existence—unhappiness, guilt, and despair.[40] For Jonas, those who deny traces of mind to lower life forms are not making an error of fact but are expressing a metaphysical prejudice, a form of mechanistic or reductionist materialism that is not a neutral description of the physical world.[41] In response to this modern nihilism, Jonas asserted that there is purpose and meaning in the world.

Jonas maintained that modern nihilism denied four tenets basic to the biblical/Jewish tradition: God created the universe, the universe is good, humanity is created in the divine image, and God makes known to humanity what is good.[42] Although Jonas argued that theology was not necessary to confute modern nihilism, he believed that only a theological response could answer certain basic human needs.[43] We want to know that there is a loving

39. Jonas, "Evolution and Freedom: On the Continuity among Life-Forms," in *Mortality and Morality*, 66–67.

40. Ibid., 70–74; Jonas, "The Burden and Blessing of Mortality," in *Mortality and Morality*, 88–92; Jonas, *Phenomenon of Life*, 186.

41. Lawrence Vogel, editor's introduction to *Mortality and Morality*, 11.

42. Jonas, "Contemporary Problems in Ethics from a Jewish Perspective," 258.

43. Vogel, editor's introduction, 36.

God who created the universe and still sustains it. We hope that goodness is not lost or forgotten in this universe; in some way we are inscribed in what the Jewish tradition calls the "Book of Life." Only through theology can these longings be addressed. Jonas tried to create a rational theology that was congruent with his existentialism, his metaphysics, and his understanding of modern science. But in the process of creating that theology, he had to create a model of divine action that would also be relevant to the world after the searing experience of Auschwitz, where his mother had died.

In 1987 Jonas published an essay entitled "The Concept of God after Auschwitz: A Jewish Voice," which was a revision of an earlier essay on immortality.[44] For Jonas, as soon as we begin to consider the concept of God, we are immediately confronted with Auschwitz. He believed that no previous ideas of theodicy can survive the Holocaust. At Auschwitz, "Not fidelity or infidelity, belief or unbelief, not guilt or punishment, not trial, witness and messianic hope, nay, not even strength or weakness, heroism or cowardice, defiance or submission had a place there."[45] And so, Jonas asked, "What God could let it happen?" Auschwitz calls into question the central belief in a God of history who makes the world a "locus of divine creation, justice and redemption." So unless one is willing to give up the idea of God, Jonas argued that the "Lord of history . . . will have to go by the board."[46]

Jonas then repeated a "myth" that he created for his article on immortality:

> [T]he ground of being, or the divine chose to give itself over to the chance and risk and endless variety of becoming. And wholly so: entering into the adventure of space and time, the deity held nothing back of itself: no uncommitted or unimpaired part removed to direct, correct, and ultimately guarantee the devious working-out of its destiny in creation.[47]

Jonas elaborated on three critical characteristics of this mythology. First, God is a *suffering* God. Divine suffering is the pain that God feels alongside the pain of God's creations as well as the disappointments with humanity that God experiences. As William Kaufman has put it, in his discussion of Jonas's concept of God, "Creation is tragic."[48] Second, God is a *becoming* God: "It is a

44. Now printed in Jonas, "The Concept of God after Auschwitz: A Jewish Voice," in *Mortality and Morality*, 131–43. This version is a translation of a lecture given in Germany in 1984. It was a revised version of his "The Concept of God after Auschwitz," published in *Out of the Whirlwind: A Reader of Holocaust Literature*, ed. Albert H. Friedlander (New York: Union of American Hebrew Congregations, 1968), 465–76.

45. Jonas, "The Concept of God after Auschwitz," 133.

46. Ibid.

47. Ibid., 134.

48. William E. Kaufman, *Evolving God in Jewish Process Theology* (Lewiston, NY: Mellen, 1997), 151.

God emerging in time instead of possessing a completed being that remains identical with itself throughout eternity."[49] The becoming God is affected and altered by the events occurring in the universe. If God has any relation to creation, then God must change through the process of the emerging universe. Lastly, God is a *caring* God. "Whatever the 'primordial' condition of the Godhead, he ceased to be self-contained once he let himself in for the existence of a world by creating such a world or letting it come to be." This does not mean that God intervenes in history in the traditional sense. Instead, God has left his human creations with the responsibility of acting for the sake of the universe. Therefore, God has "made his care dependent on them." This implies that God is also an "endangered God, a God who runs a risk."[50]

The implication of this analysis is that God is not omnipotent: "for the sake of our image of God and our whole relation to the divine, for the sake of any viable theology, we cannot uphold the time-honored (medieval) doctrine of absolute, unlimited divine power."[51] Jonas advanced two additional arguments for the rejection of divine omnipotence. First, from a purely logical perspective, omnipotence is self-contradictory since absolute power implies that there can be no other object on which to exercise its power. The existence of such an other implies limitation. An omnipotent God, therefore, could only exist in solitude, with nothing on which to act.[52] Second, and most importantly for Jonas, there is a theological argument against divine omnipotence: unless God is completely inscrutable, absolute power can exist in God only at the expense of absolute goodness: "Only a completely unintelligible God can be said to be absolutely good and absolutely powerful, yet tolerate the world as it is."[53] Jonas characterized God as good and intelligible, but not omnipotent. He believed that after Auschwitz, goodness can be maintained only at the cost of relinquishing power. Therefore God was silent at Auschwitz, not because God chose not to intervene, but because God could not intervene. Rejecting a Manichean dualism to explain the origin of evil, Jonas located evil in the deliberate acts of human beings.[54]

Thus God's limitation is a self-limitation sacrificed at the moment of Creation. Here Jonas connected his idea of divinity with the kabbalistic doctrine of *tzimtzum*, the contraction of the divine for the purpose of

49. Jonas, "The Concept of God after Auschwitz," 137.

50. Ibid., 138.

51. Ibid.

52. Ibid., 138–39. See also Charles Hartshorne, *Omnipotence and Other Theological Mistakes* (Albany: State University of New York Press, 1984).

53. Jonas, "The Concept of God after Auschwitz," 139.

54. Ibid., 141.

creation. For Jonas, however, the contraction is total: "the Infinite ceded his power to the finite and thereby wholly delivered his cause into its hands."[55] What is then left of the relation of creation to its Creator? Jonas believed that "[h]aving given himself whole to the becoming world, God has no more to give: it is man's now to give to him."[56] In a later essay, Jonas retreated somewhat from this idea and suggested that God continues to act in relation to the universe.[57] God does not do this through direct supernatural actions, which would contradict the laws of the natural world, but rather through the inspiration of certain individuals. If our freedom to act in the world is scientifically compatible with causality, then we can accept a kind of divine causality that comes into our inner self, but does not conflict with human free will.[58]

Jonas's model of divine action was not only developed from his philosophy of the organism, but also from his moral stance on the necessity for an intelligible theology after the Holocaust. His concept of creation is similar to Davies'—divine action consists only in an initial creation—but with a much larger sense of divine risk. While he leaves some room for revelation, he never fully explains what he means by divine inspiration. In Jonas's theology, God feels all that happens in the universe, is affected by all that occurs, and preserves all "perishing occasions everlastingly."[59] But while there is no loss, there is redemption only within God. Jonas does not allow for final cosmic or human redemption. In fact, he rejects Teilhard de Chardin's idea that the universe is growing inevitably towards an Omega point.[60] For Jonas, the future of the universe is still a risk for God and the outcome uncertain.

John Haught

The Catholic theologian John F. Haught in *God after Darwin: A Theology of Evolution*, has written one of the most detailed analyses of the relationship of theology to evolution. He claims that "[a]fter Darwin, we may still think of God as powerfully effective in the natural world, but we will have to do so in a manner quite distinct from that implied in much pre-evolutionary theology."[61]

55. Ibid., 142.

56. Ibid.

57. Jonas, "Is Faith Still Possible?: Memories of Rudolf Bultmann and Reflections on the Philosophical Aspects of His Work," in *Mortality and Morality*, 154 f.

58. Ibid., 156–60.

59. Haught, *God after Darwin*, 184.

60. Jonas, "Matter, Mind, and Creation: Cosmological Evidence and Cosmogonic Speculation," in *Mortality and Morality*, 188 f.

61. Haught, *God after Darwin*, 46.

He believes that the story of life, even from a neo-Darwinist perspective, provides us with essential concepts for thinking about God and God's relation to humanity and the natural world. Our new awareness of cosmic and biological evolution should enhance and enrich the traditional teachings about God and God's way of acting in the world, especially in regards to creation, eschatology, revelation, divine love (or grace), divine power, and redemption.[62]

While the traditional notion of creation consisted solely of the original creation, evolution allows theology to understand that creation is not only an original act, but also an ongoing and constantly new reality. Had creation been completed at the beginning, the world would be without a future and devoid of life; the universe is unfinished and in such a universe perfection is impossible and evil is inevitable. In Haught's evolutionary theology, revelation is not "fundamentally the communication of propositional information from a divine source of information. Rather, it is at root the communication of *God's own selfhood* to the world."[63] Moreover, evolution suggests that if God loves the world, then God's grace means letting the world be itself and "refrain[ing] from forcefully stamping the divine presence or will upon the world or dissolving the world into God." Divine love must be in the form of self-withdrawal in order for the world to emerge on its own.[64]

Haught finds that process theology offers the best way to define divine action in an evolving universe. It interprets the biblical concept of God's creative and redemptive action in a way that coheres with the dynamic evolutionary character of the world. Process theology attributes evolution to divine action, which takes the form of persuasive love rather than coercive force. God invites and does not compel. This is necessary both for human freedom and for the novelty required in creation for the universe to be something other than its Creator: "[that] God's creation is not driven coercively, that it is widely experimental, and that it unfolds over a considerable amount of time."[65] Process theology can thus incorporate quantum indeterminacy and undirected genetic mutation into its theological framework of divine action. God is the source not only of order but also of novelty, which can take the form of instability and disorder and makes evolution possible.[66] Evolution is the world's response to God's desire that it struggle toward ever more complex patterns of existence.

62. Ibid., 36–43.

63. Ibid., 38. Italics in the original.

64. Ibid., 40.

65. Ibid., 42.

66. Ibid., 41–43.

Haught also uses process theology to understand the concept of redemption in an evolving universe, where constant death is part of the process. In process theology, God is infinitely responsive to the world, "feels," and is influenced by all that happens during the process of evolution. Everything that happens through evolution is preserved by being taken eternally into God's own feeling of the world. Thus each event is redeemed from death and thus has ultimate meaning, although we cannot see this clearly because we live in an unfinished universe.[67] Haught also believes that the evolving universe has the character of promise, an eschatological category in which there will be a complete unfolding of God's plan in the future:

> The thrust of much recent science, and especially evolution, is that we truly belong to the universe. Theologically this would mean that the revelatory promise that gives us our hope extends backward to cosmic beginnings, outward to the most remote galaxies, and forward to the future of the whole creation. And if all of nature shares in the promise, then this should be more than enough reason for taking care of it here and now as we wait "in joyful hope" for its fulfilment in God's new creation.[68]

Haught, like Jonas, accepts the notion of a "kenotic" or self-emptying God, which he believes fits the picture of the universe painted by evolutionary science. He places the image of a self-emptying God at heart of the Christian concept of revelation and the doctrine of the Trinity. This idea of divinity also allows for God's other attributes, such as his power, covenant, redemption, justice, or wisdom, but joins them to the idea of divine vulnerability. This concept of a "kenotic" God also brings new meaning to the suffering and loss that is found in life. The self-withdrawal of God makes not only the initial Creation possible, but also ongoing creation. Ongoing creation through evolution is the world responding to divine "allurement" at its own pace and in its own particular way, being self-ordering and self-creative during immense amounts of time. Therefore, says Haught, divine humility is a reasonable metaphysical explanation of the evolutionary process as seen in modern science. All scientific explanations must lie within some metaphysical vision of reality and this one, Haught believes, is superior in explanatory power to a materialistic metaphysics.[69]

The self-withdrawal of God also allows for the future to happen. Following Teilhard, Haught argues that a metaphysics of the future explains

67. Ibid., 43.
68. Ibid., 164.
69. Ibid., 45–55.

the natural world's contingency, lawfulness, and temporality. Contingency is a sign of nature's openness to new creation. Lawfulness is essential for the emergence of novelty. Novelty cannot be planted in chaos, but needs previous forms to overcome. Nature also needs some degree of regularity and predictability in order for the world to remain consistent enough to have a future. Temporality means that the arrival of the future "allows each present to retreat irreversibly into the fixed past so that other new moments may arise in its place."[70]

Haught asks how we can understand God's influence on the universe, especially before the emergence of life. If inanimate matter can become life, mind, and spirit, there must be some properties of the physical universe that allow it to be drawn into these patterns. If physical reality is mindless, how is it able to respond to the divine power of attraction? What hidden characteristics did the universe have that allowed it to be moved by God and to create life? Process theology extends "subjectivity" and therefore responsiveness to the basic components of matter, which would then allow them to respond to God. For materialists, subjectivity is an epiphenomenon and mind has no essential place in nature. If this is so, then our choice is either to ignore the emergent reality of mental interiority or to "squeeze it into categories shaped by a method of inquiry that has decided from the start to abstract from it altogether."[71] Haught rejects this approach and, agreeing with Jonas, believes that subjectivity is an objective fact of the universe.[72] However, he follows Whitehead in claiming that subjectivity existed before the beginning of life. With the appearance of life in cosmic evolution, potential interiority becomes actual interiority. And while there is a sharp distinction between the two, nonetheless the potential for life before life suggests that mind has always been entangled with matter.[73]

Haught has created a model of divine action that fits with evolution as he understands it and is able to preserve the doctrines of creation, revelation, and redemption from a Catholic perspective. While his terminology is Christian, many of his observations and deductions can also be viewed through the lens of Jewish theology in order to create a new model of divine action and providence.

70. Ibid., 103.

71. Ibid., 168. In this, Haught follows Michael Polanyi, *Personal Knowledge: Toward a Post-Critical Philosophy* (Chicago: University of Chicago Press, 1962).

72. Jonas, "Matter, Mind, and Creation," 165–97.

73. Haught, *God after Darwin*, 180–84.

Towards a New Jewish Theology of Divine Action

One implication of Haught's writings is that theological reflection can be enhanced by drawing on scientific theories, such as evolution. This is not to effect a facile reconciliation between science and religion, but rather to enrich our theological categories. This is also the approach of Jonas, who emphasizes the moral necessity to change our view of God's relation to the world. Such a theology can reinvigorate our concepts of creation, revelation, and redemption by showing how God can still be active in the world while preserving human freedom and at the same time not violating the laws of nature. A new Jewish theology of divine action in the light of evolution will provide an alternative to the classical dualism of spirit and matter and create a kind of neo-naturalism, which is more consonant with biblical thought. Following the perspectives of Davies, Jonas, and Haught, such a neo-naturalism preserves the significance of human consciousness, which is essential in a Jewish theology committed to portraying humanity as created in the image of God. An evolutionary theology can help us to expand and enrich several contemporary Jewish concepts of creation, revelation, and redemption.

Many Jewish theologians speak of a limited or self-limiting God for many of the same reasons that Haught suggests. For example, philosopher William Kaufman has shown how process theology has influenced Mordecai Kaplan, Abraham Joshua Heschel, and other modern Jewish thinkers.[74] An evolutionary Jewish theology of divine action would include a notion of divine limitation, thus preserving human freedom and offering another approach to the problem of evil. It would also take into account the evolutionary character of the whole universe, whose development requires both contingency and law. Human freedom would be understood as part of a larger process of ongoing creation that is constantly producing novelty, but also requiring destruction for further regeneration. God began this process and is part of it but does not violate the laws of nature in continuous divine interaction. With evolutionary theology, creation can become again more than a symbolic term. It can restore a sense of meaning and purpose to the universe.

A new Jewish theology of divine action must also account for revelation. To achieve this within a naturalistic view of the universe, many modern Jewish theories of revelation share with Haught the idea of God's presence being continuously revealed. Sol Tanenzapf, for example, has shown how the concepts of the revelation of Torah and mitzvot (good deeds) can be seen

74. See Kaufman, *Evolving God.* See also the various Jewish process theology positions in Sandra B. Lubarsky and David Ray Griffin, eds., *Jewish Theology and Process Thought* (Albany: State University of New York Press, 1996).

from a process perspective.[75] This coincides with the views of Elliot Dorff and Norbert Samuelson, who argue that most modern Jewish theologies redefine revelation as an encounter with God's own self rather than as the communication of specific laws and beliefs.[76] Thus, God can continue to influence the world through inspiration and the encounter with the divine presence. Evolutionary theology does not exclude this kind of revelation, as it does not violate the laws of nature. But evolutionary theology can also enrich the concept of revelation if we understand that the universe itself provides us with another source of revelation. By understanding how the laws of nature work, something of the nature of God is revealed that perhaps can lead us to a new kind of creation spirituality or Jewish environmental ethics.[77]

Finally, a Jewish theology of divine action must include some concept of redemption. Many modern Jewish theologies of redemption could encompass a model similar to Haught's concept of promise. For example, Norbert Samuelson has suggested that a Jewish theology of creation requires a corresponding concept of redemption: "the universe originates in the nothing of creation as a process towards the something of redemption."[78] Even though Samuelson's "something of redemption" is outside the purview of science in that "redemption is an ethical utopian concept,"[79] he too speaks about how creation holds out a promise. In his view the promise is that the universe is becoming good; not a human good but a divine good. Likewise in a review of modern Jewish concepts of eschatology, Neil Gillman has shown how they

75. Sol Tanenzapf, "A Process Theory of Torah and Mitzvot," ibid., 35–46.

76. Elliot Dorff, "Medieval and Modern Theories of Revelation," in *Etz Hayim: Torah and Commentary*, ed. David Lieber et al. (New York: Rabbinical Assembly and United Synagogue of Conservative Judaism, 2001), 1401–5; Norbert M. Samuelson, *Revelation and the God of Israel* (Cambridge: Cambridge University Press, 2002).

77. One possible theological framework for this new Jewish creation theology would be to develop a Jewish version of the "two books of God" metaphor. Christian theologians and scientists in the sixteenth to eighteenth centuries often used a theological construct called the "two books of God," the Book of Nature and the Book of Scripture, to describe the relationship between scripture and the creation. For the origin of this concept see Ernst Robert Curtius, *European Literature and the Latin Middle Ages* (Princeton: Princeton University Press, 1990 [1953]), 315–26. Menachem Fisch (private communication) notes in an unpublished lecture that the "two books" metaphor was not used by Jewish writers. I have recently considered how this metaphor may be used in Jewish theology. See Lawrence Troster, "The Two Books of God," *Sh'ma*, December 2005, 4–5.

78. Norbert M. Samuelson, *Judaism and the Doctrine of Creation* (Cambridge: Cambridge University Press, 1994), 240.

79. Ibid.

tend to view the traditional idea of resurrection as a symbol that points towards a future hope beyond time and space.[80] Haught's idea that God preserves everything that happens suggests the notion of an afterlife that redeems every event from death and creates ultimate meaning. Jonas had a similar concept in his idea that minimally God feels all that happens in the universe and preserves all events eternally. The classical Jewish concept of redemption with its "death of death" and the coming of the Messiah may be the most difficult to connect with an evolutionary theology.[81] Classic Jewish redemption is precisely about the transcending of the natural world and the miraculous return to an Eden-like existence in which life is completely triumphant.[82]

While the scientific challenge of evolution forces us to confront our ideas of divine action and of providence, we can nonetheless endorse Haught's sentiment that this challenge is Darwin's gift to theology: "Evolutionary biology not only allows theology to enlarge its sense of God's creativity by extending it over measureless eons of time; it also gives comparable magnitude to our sense of the divine participation in life's long and often tormented journey."[83]

80. Gillman, *Death of Death*, 215–41.

81. I do not believe that it is unsolvable, and perhaps it is possible to create an evolutionary eschatology from an adaptation of the kabbalistic idea of the "other side." I hope to deal with this in a future article.

82. This is found in the classic vision of the messianic era in Isaiah 11.

83. Haught, *God after Darwin*, 46.

Suggested Reading

Aviezer, Nathan. *Fossils and Faith: Understanding Torah and Science.* Hoboken, NJ: Ktav, 2001.

———. *In the Beginning . . . : Biblical Creation and Science.* Hoboken, NJ: Ktav, 1990.

Barkan, Elazar. *The Retreat of Scientific Racism: Changing Concepts of Race in Britain and the United States between the World Wars.* Cambridge: Cambridge University Press, 1992.

Branover, Herman, and Ilana Coven Attia, eds. *Science in the Light of Torah: A B'or Ha'Torah Reader.* Northvale, NJ: Aronson, 1994.

Cantor, Geoffrey. *Quakers, Jews, and Science: Religious Responses to Modernity and the Sciences in Britain, 1650–1900.* Oxford: Oxford University Press, 2005.

Carmell, Aryeh, and Cyril Domb, eds. *Challenge: Torah Views on Science and Its Problems.* London: Association of Orthodox Jewish Scientists; Jerusalem: Feldheim, 1976.

Cherry, Michael Shai. "Creation, Evolution, and Jewish Thought." Ph.D. dissertation, Brandeis University, 2001.

———. "Three Twentieth-Century Jewish Responses to Evolutionary Theory." *Aleph: Historical Studies in Science and Judaism* 3 (2003): 247–90.

Cohen, Naomi W. "The Challenges of Darwinism and Biblical Criticism to American Judaism." *Modern Judaism* 4 (1984): 121–57.

Dodson, Edward O. "*Toldot Adam*: A Little-Known Chapter in the History of Darwinism." *Perspectives on Science and Christian Faith* 52 (2000): 47–54.

Dubin, Lois. "*Pe'er ha-Adam* of Vittorio Ḥayim Castiglioni: An Italian Chapter in the History of Jewish Response to Darwin." In *The Interaction of Scientific and Jewish Cultures in Modern Times,* ed. Yakov Rabkin and Ira Robinson. Lewiston, NY: Mellen, 1995, 87–102.

Efron, John M. *Defenders of the Race: Jewish Doctors and Race Science in Fin-de-Siècle Europe.* New Haven: Yale University Press, 1994.

Falk, Raphael. "Zionism and the Biology of the Jews." *Science in Context* 11 (1998): 587–607.

Faur, José. "The Hebrew Species Concept and the Origin of Evolution: R. Benamozegh's Response to Darwin." *Rassegna Mensile di Israel* 63 (1997): 42–66.

Feit, Carl. "Darwin and Drash: The Interplay of Torah and Biology." *Torah u-Madda Journal* 2 (1990): 25–36.

Gasman, Daniel. *Haeckel's Monism and the Birth of Fascist Ideology.* New York: Lang, 1998.

Gilman, Sander. *The Jew's Body.* New York: Routledge, 1981.

Goldfarb, Stephen J. "American Judaism and the Scopes Trial." In *Studies in the American Jewish Experience II*, ed. Jacob R. Marcus and Abraham J. Peck. Lanham, MD: University Press of America, 1981, 33–47.

Green, Arthur. *Ehyeh: A Kabbalah for Tomorrow*. Woodstock, VT: Jewish Lights, 2003.

Hart, Mitchell. "Racial Science, Social Science, and the Politics of Jewish Assimilation." *Isis* 90 (1999): 268–97.

———. *Social Science and the Politics of Modern Jewish Identity*. Stanford: Stanford University Press, 2000.

Kaplan, Lawrence. "*Torah u-Madda* in the Thought of Rabbi Samson Raphael Hirsch." *Bekhol Derakhekha Daehu* 5 (1997): 5–31.

Kirsh, Nurit. "Population Genetics in Israel in the 1950s: The Unconscious Internalization of Ideology." *Isis* 94 (2003): 631–55.

Kohn, David, and Ralph Colp. " 'A Real Curiosity': Charles Darwin Reflects on a Communication from Rabbi Naphtali Levy." *European Legacy* 1 (1996): 1716–27.

Landa, Judah. *Torah and Science*. Hoboken, NJ: Ktav, 1991.

Lilienthal, Georg. "Die jüdischen 'Rassenmerkmale': Zur Geschichte der Anthropologie der Juden." *Medizinhistorisches Journal* 28 (1993): 173–98.

Matt, Daniel. *God and the Big Bang: Discovering the Harmony between Science and Spirituality*. Woodstock, VT: Jewish Lights, 1996.

Meirovich, Harvey Warren. *The Vindication of Judaism: The Polemics of the Hertz Pentateuch*. New York: Jewish Theological Seminary of America, 1998.

Mocek, Reinhard. *Biologie und soziale Befreiung: Zur Geschichte des Biologismus und der Rassenhygiene in der Arbeiterbewegung*. Frankfurt: Lang, 2002.

Mosse, George L. *Toward the Final Solution: A History of European Racism*. New York: Fertig, 1978.

Müller-Hill, Benno. *Murderous Science: Elimination by Scientific Selection of Jews, Gypsies, and Others in Germany, 1933–1945*. Trans. George R. Fraser. Oxford: Oxford University Press, 1988.

Nussbaum, Alexander. "Creationism and Geocentrism among Orthodox Jewish Scientists." *National Center for Science Education Reports* 22 no. 2 (2002): 38–43.

Poliakov, Leon. *The Aryan Myth: A History of Racist and Nationalist Ideas in Europe*. London: Sussex University Press, 1974.

Proctor, Robert. *Racial Hygiene: Medicine under the Nazis*. Cambridge: Cambridge University Press, 1988.

Robinson, Ira. "Judaism since 1700." In *The History of Science and Religion in the Western Tradition: An Encyclopedia*, ed. Gary B. Ferngren et al. New York and London: Garland, 2000, 288–90.

Rosenberg, Shalom. "Introduction to the Thought of Rav Kook." In *The World of Rav Kook's Thought*, ed. Benjamin Ish-Shalom and Shalom Rosenberg, trans. Shalom Carmy and Bernard Casper. Jerusalem: Avi Chai, 1991, 88–97.

Samuelson, Norbert M. *Revelation and the God of Israel*. Cambridge: Cambridge University Press, 2002.

Sarna, Nahum M. "Understanding Creation in Genesis." In *Is God a Creationist? The Religious Case against Creation-Science*, ed. Roland Mushat Frye. New York: Scribners, 1983, 155–75.

Schneersohn, Menachem Mendel. *Mind Over Matter: The Lubavitcher Rebbe on Science, Technology, and Medicine*. Ed. and trans. Arnie Gotfryd. Jerusalem: Shamir, 2003.

Schroeder, Gerald. *Genesis and the Big Bang: The Discovery of Harmony between Modern Science and the Bible.* New York: Bantam, 1990.

———. *The Hidden Face of God: How Science Reveals the Ultimate Truth.* New York: Free Press, 2001.

———. *The Science of God: The Convergence of Scientific and Biblical Wisdom.* New York: Free Press, 1997.

Scult, Mel. *Judaism Faces the Twentieth Century: A Biography of Mordecai M. Kaplan.* Detroit: Wayne State University Press, 1993.

Slifkin, Nosson. *The Science of Torah: The Reflection of Torah in the Laws of Science, the Creation of the Universe, and the Development of Life.* Southfield, MI: Targum; Nanuet, NY: Feldheim, 2001.

Staub, Jacob J. "Evolving Definitions of Evolution." *Reconstructionist* 61 no. 2 (1996): 4–13.

Sterman, Baruch. "Judaism and Darwinian Evolution." *Tradition* 29 (1994): 48–75.

Swetlitz, Marc. "Responses of American Reform Rabbis to Evolutionary Theory, 1864–1888." In *The Interaction of Scientific and Jewish Cultures in Modern Times,* ed. Yakov Rabkin and Ira Robinson. Lewiston, NY: Mellen, 1995, 103–25.

———. "American Jewish Responses to Darwin and Evolutionary Theory, 1860–1890." In *Disseminating Darwinism: The Role of Place, Race, Religion, and Gender,* ed. Ronald L. Numbers and John Stenhouse. Cambridge: Cambridge University Press, 1999, 209–46.

Tirosh-Samuelson, Hava. "Judaism." In *Encyclopedia of Science and Religion,* 2 vols., ed. J. Wentzel van Huyssteen. New York: Macmillan, 2003, 1:477–83.

Weikart, Richard. *From Darwin to Hitler: Evolutionary Ethics, Eugenics, and Racism in Germany.* New York: Palgrave Macmillan, 2004.

Weindling, Paul. *Health, Race, and German Politics between National Unification and Nazism, 1870–1945.* Cambridge: Cambridge University Press, 1989.

Wolowelsky, Joel. "Teaching Evolution in Yeshiva High School." *Ten Da'at: A Journal of Jewish Education* 10 no. 1 (spring 1997): 33–39. Available online at http://www.daat.ac.il/daat/english/education/evolution-1.htm.

Index